MINERALOGY—APPLICATIONS TO THE MINERALS INDUSTRY

PAUL F. KERR — Age 72

Newberry Professor Emeritus
Columbia University, New York

MINERALOGY—APPLICATIONS TO THE MINERALS INDUSTRY

Proceedings of the
Paul F. Kerr Memorial Symposium
New York, New York

February 28, 1985

Edited by
Donald M. Hausen
Newmont Exploration Limited
Danbury, Connecticut
and
Otto C. Kopp
Department of Geological Sciences
University of Tennessee
Knoxville, Tennessee

Published by the
Society of Mining Engineers
of the
American Institute of Mining, Metallurgical,
and Petroleum Engineers, Inc.
New York, New York • 1985

Printed in the United States of America
by BookCrafters, Inc., Chelsea, Michigan

Library of Congress Catalog Card Number 85-71781
ISBN 0-89520-446-0

FOREWORD

This volume contains the proceedings of 16 papers and abstracts that were presented during the two sessions of the *Dr. Paul F. Kerr Memorial Symposium*, held on February 28, 1985, at the SME-TMS-AIME 114th Annual Meeting in New York City. The papers delve into a variety of specialized topics in applied mineralogy, notably those of singular interest to Dr. Kerr during his most productive years as a university professor and thesis advisor to many doctoral candidates. Topics range from uranium, vanadium and clay mineral studies, to geochemistry, alteration and ore microscopy, with special emphasis on innovations in methodology.

This collection of timely research papers serves as a personal tribute to "Pappy" Kerr and his role as Father of Applied Mineralogy in this country. Former students, co-workers, friends and family members gathered together to participate in this memorial symposium to convey their love and respect for this man and to share their personal reminiscences of his impact on their careers. Participation was deemed a privileged opportunity to express their gratitude for his personal guidance and friendship at a crucial time in their lives...when they were graduate students.

Dr. Kerr was one of the earliest pioneers of "process" mineralogy in the United States, where he constructed one of the first x-ray diffraction units. He was not satisfied with just the physical academics of x-rays and crystal structure, per se, and started applying XRD methodology right away to a variety of industrial and geologic problems in the early 1920s. As an educator, he was successful in communicating his ideas and innate curiosity in nature's systems to his progeny of graduate students and colleagues. This influence extended well into his twilight years as Newberry Professor Emeritus at Columbia University in the late 1960s and early 1970s.

The proximity of this Symposium in New York City to the halls of Columbia University provided a time and place to meet once again at the Faculty Club on Morningside Heights for a testimonial banquet on Thursday evening, to commemorate his passing four years ago on February 27, 1981. It was here Dr. Kerr had hosted and encouraged his graduate students individually on countless former occasions.

The Editors take special pride in dedicating this first Kerr Volume to his memory, and look forward to future memorial symposia, where his colleagues and friends, first, second, third generations and beyond, may convene and discuss new methodology and mineralogic applications in the elegant and enthusiastic style of our late mentor, Dr. Paul F. "Pappy" Kerr.

D.M. Hausen
O.C. Kopp
Editors

ACKNOWLEDGMENTS

The manuscripts were reviewed by the following people, representing a diversity of specialized fields in mineralogy, geology, and engineering. We would like to thank each of them for their prompt and effective reviews of one or more papers contained in this symposium volume.

J.W. Ahlrichs
P.B. Barton, Jr.
W.A. Bassett
P.M. Bethke
M.W. Bodine
E.N. Cameron
J.A. Coope
W.J. Croft
H.M. Dahl
R.D. Hagni
W.C. Hellyer
K.S. Johnson
P.L. Kistler
J.R. Kyle
B.F. Leonard
M.W. Malloy
E.L. Ohle
W.C. Park
W. Petruk
E.M. Ripley
A.H. Vassiliou

The editors wish to acknowledge the special efforts of Andy Vassiliou for his organization of the testimonial dinner at the Columbia University Faculty Club in honor of Dr. Kerr, and Phil Bethke for his general assistance in organizing the symposium.

The management of Newmont Exploration Limited and the University of Tennessee Department of Geological Sciences gave full support for the editorial tasks. Special appreciation is extended to Pauline Kistler for her careful editing, organizing, and typing of manuscripts, and to Karen Beck for her efforts in correspondence and typing of papers. Acknowledgments are also given to Paul Barton, Jr. and his secretary, Rosie Gee, for their assistance in editing and typing of manuscripts.

TABLE OF CONTENTS

MEMORIALS AND TESTIMONIALS

KEYNOTE ADDRESS

Chapter 1

DR. KERR'S CONTRIBUTION TO ORE DISCOVERY

by Leo J. Miller

Miller Exploration Company
Evergreen, CO 80439

It is both an honor and a pleasure for me to be the keynote speaker for the Paul F. Kerr's Symposium on Applied Mineralogy. Professor Kerr was a man for all seasons to his students, and he treated all students as distinct individuals. Philip Bethke in his memorial to Dr. Kerr summed up our remembrance best by stating that: "We have lost our most successful educator-administrator-scholar, who was our steadfast teacher-counselor-mentor-friend." His unique trait was the ability to bring out the best in a variety of individuals in order that their performance was at the peak of quality.

A Biography

Paul Kerr was born in Hemet, California, on January 12, 1897. He graduated from Occidental College in 1919 and received his Ph.D. from Stanford in 1923. While at Stanford, he constructed a multiple x-ray diffraction instrument which was the basis of his doctoral dissertation, "The Determination of Opaque Ore Minerals by X-Ray Diffraction Patterns." He was one of the early pioneers in the technique of powder diffraction, and he utilized this tool in most of his work throughout his life.

Kerr joined the Stanford faculty during the fall term of 1923 as a substitute teacher for Professor A. F. Rogers. In the spring of 1924 he grasped at an opportunity which was to establish his scholarly career, and, in due course, to affect the lives of all of us. It was at this time he moved to Columbia University for an association which lasted 41 years. Dr. Kerr started at Columbia as a temporary lecturer and in a short time became a permanent member of the faculty. He was a full professor by 1940 and Newberry Professor of

Mineralogy in 1959. Dr. Kerr retired from Columbia University in 1965 and moved back to California as consulting professor at Stanford until 1977. On February 26, 1981, Paul Kerr was to be the keynote speaker for the American Institute of Mining, Metallurgy and Petroleum Engineers Symposium on "Process Mineralogy: Extractive Metallurgy, Mineral Exploration, Energy Resources." Don Hausen organized this symposium as an honor to Kerr and requested former students, such as myself, to present papers. On that fateful morning in Chicago we were notified that Paul Kerr had fallen ill, hopefully of the flu. However, his illness was more serious, and on the next day we lost our dear friend through a heart attack. Dr. Hausen read Kerr's paper entitled, "Reminiscences in Applied Mineralogy," which represents a true insight into the delightful character of Paul Kerr.

Achievements

In addition to his half century of teaching, Paul Kerr was an especially effective researcher in mineralogy since he consistently wedded his basic research to application. He excelled in the mineralogy of clays and metals, and at the same time developed the tools necessary for identification. In the early 1930's, Kerr collaborated with Ross to establish the first modern classification of clay minerals. He utilized this background to better understand the properties of quickclays and the clay alteration surrounding certain types of ore deposits. Kerr maintained that mapping various clay species could lead to predictions on the timing of landslides, slope failure, and earth dam collapse. I can clearly remember the discussions between Sidney Page, lecturer in engineering geology, and Paul Kerr during 1953 and 1954 concerning the failure of the Fort Peck dam on the Missouri River. The Fort Peck dam was the first hydraulic clay core dam in the world. It failed catastrophically one evening, and Page concluded it was due to piping, while Kerr favored bentonitic swelling at the base. Both men presented their facts well, but Kerr's knowledge of quickclays won the day.

Paul Kerr's interest in ore deposits commenced with his doctoral thesis and continued throughout his life. His greatest contributions concerned tungsten, porphyry copper, and uranium deposits. His work with gemstones, dumortierite (spark plug minerals) and saline deposits has also been of great value. Between 1950 and 1970 Kerr completed studies on the clay minerals of the Gulf Coast deltaic sediments, and ascertained the reservoir potential of certain basins.

Ore Discovery

Paul Kerr had the unique ability to deal with students with a wide variety of interests. His students ranged from basic scientists to business men, and all the shades in between. The main purpose of this presentation is to discuss Kerr's contribution to ore discoveries made by men that he influenced in applied sciences.

In the early 1930's Paul Kerr taught a student by the name of Phil Jenney. Phil obtained his Ph.D. at Columbia under his mentor, Kerr, with his dissertation on the spark plug minerals. I met Phil in 1955 in Bathurst, New Brunswick, during the beginning of the era of massive-sulphide exploration dominance. Phil had been instrumental in discovering the Heath Steel Mine, and we discussed our favorite professor many times during the dreary New Brunswick winters. He had come to Paul Kerr in 1933 torn between a desire to be rich and a scientist. Kerr taught him the compatability of both, and therefore nurtured for our civilization a much needed blend of scientist-promoter-explorer. Jenney was an adherent to Kerr's concept that ore controls in similar ore systems should not change. Armed with this conviction and his success in New Brunswick, Jenney organized a group sponsored by the Canadian promoter, Carl Springer, to explore for massive sulphide deposits in the Canadian Shield. He chose an area near Mattagami Lake, and named his group the Mattagami Syndicate.

On the fringe of a large anorthositic complex there existed several magnetic anomalies as at New Brunswick. However, the central complex is highly magnetic, while at New Brunswick it is composed of magnetically low rhyolitic ash flows. In any event, Phil Jenney commenced an aerial electromagnetic survey along the fringe of the magnetic complex over small magnetic outlying anomalies. He discovered the first hidden massive sulphide deposit on the Canadian Shield. This was the beginning of a new 30 year era of Canadian exploration, which is now ending.

Paul Kerr was one of the first American geologists to become interested in uranium deposits. During the years of the Manhattan Project, Kerr traveled to the Belgian Congo to investigate the Shinkolobwe copper-uranium mine. Upon his recommendation and that of his former student, Phil Merritt, the mine was reopened for uranium production. Phil Merritt became head of uranium procurement for the Atomic Energy Commission, and by 1952 was requesting Kerr's advice in numerous areas of uranium exploration. It was during this period of time that Paul Kerr arrived at our A.E.C. camp in Fry Canyon on the Colorado Plateau. He had just completed his clay alteration studies at Santa Rita and Silver Bell, and therefore was established as a true leader in the field of metal exploration. In the evening, around a campfire, Paul Kerr discussed the hydrothermal alteration surrounding porphyry copper systems. We had all studied the work of Sales and Meyer at Butte, and heard of the discoveries of Richards and Courtwright in the southwest, but now we were learning the details of alteration first hand. One of the men involved in these discussions was a young mining engineer who desperately wished to become a geologist; his name, Dave Lowell. Following Kerr's departure, Dave went on to Kerr's alma mater, Stanford

University, and received his Master's Degree in geology. The Lowell-Guilbert model of porphyry alteration systems has won world-wide acclaim. Lowell's copper porphyry discoveries at Kalamazoo and Casa Grande West are remarkable, while his consulting clay alteration work has proven successful in South America and the Phillippine Islands. While recently discussing Paul Kerr's visits to Fry Canyon with Lowell, he reminded me of one of Kerr's humorous departures. A wet night preceded Kerr's drive to Blanding, Utah. Dave, in the morning, followed Kerr's jeep tracks for several miles and was somewhat concerned in that they slid from one side of the muddy road to the other. His fears mounted as the tracks took a sudden skid and disappeared into the pinon and juniper bush. Dave walked a few yards off the road and found that the trail ended with a wheel track on either side of a pinon tree, but the jeep was gone. We never did learn how Kerr extracted the jeep from that tree.

Paul Kerr's visits to Fry Canyon dramatically affected my career. I had just returned from a three month graduate school effort at Amherst under Professor George Bain. Dr. Bain was one of the finest ore deposit geologists of his day and, of course, a fine gentleman. In one of our many heated debates on geology, I rudely departed Amherst and returned to my previous job with the A.E.C. Paul Kerr convinced me to try again for a graduate degree, despite my poor performance at Amherst. As did Lowell, I shortly departed, except in the opposite direction...to Columbia.

While at Columbia University, Kerr fortified my belief in ore-control mapping and the building of models based on data. He warned me that if one builds a model on too little factual information, then he is a "promoter." He added that if one awaits the arrival of a large amount of data, then he is indecisive. Therefore, one must obtain the ability to act decisively the moment that sufficient information has been accumulated to find the ore deposit.

Between 1955 and 1957, I commenced to develop a volcanic theory for the origin of massive sulphide deposits. In its final form in 1957, it consisted of:

1. All massive sulphide deposits are within volcanic arcs;
2. Ninety percent are within rhyolite ash flows;
3. They are formed during the diagenic consolidation of the ash flow.

On my trips to New York, Kerr and I would discuss this model and techniques for utilization within an exploration program. During one of these meetings, I had finished outlining the various fossil volcanic arcs of the eastern Canadian Shield, and was attempting to determine which arc to spend the most exploration dollars on, and where on the arc to start exploration. Many of you who know Paul Kerr can envision him leaning back in his large chair while stroking his friendly double chin, all the while gazing at a 12 foot x 6 foot map covering 500 miles of the Shield. Paul Kerr's prophetic words were, "Select the arc which has proven massive sulphide mineralization and narrow your search to that portion of the arc which has minor mineralization. The middle arc on your map has all of these ingredients." When Kerr mentioned minor mineralization, he only meant for me to remain away from the competitive mining camps, such as Noranda. On March 13, 1959, on Texasgulf's first test flight on the Canadian Shield, I flew our electromagnetic system, mounted on a helicopter, over the strike length of what is now the Kidd Creek ore deposit.

Kerr's contribution to Kidd Creek did not cease there. While constructing maps for my doctoral dissertation, he laid down four rules which I was never to forget. Maps must always have: (1) a north arrow; (2) a scale; (3) a date; and (4) a signature. For reasons that I prefer to forget, the 1959 Kidd Creek E M anomaly was not drilled until 1963. During this interval my volcanic model lost support and the Kidd Creek anomaly was just about given away. Sometime in 1962 it was decided to option most of our Canadian land to Carl Springer, Phil Jenney's old associate. However, Texasgulf would keep certain lands that required additional work. I was working on phosphate in North Carolina when our Texasgulf Canadian group drew the option boundary line on a 1" to 2 mile map within 1/4 to 1/2 mile of the Kidd Creek anomaly. The Kidd Creek deposit was discovered in 1963, only 0.1 inch on the map from the boundary of the retained property, and Carl Springer sued Texasgulf for 250 million dollars in 1965 and 1966. Texasgulf won, primarily because each of my maps which emphasized the importance of the Kidd Creek anomaly, had Kerr's (1) north arrow, (2) scale, (3) date, and (4) signature.

Kerr's help did not end with discovery. We sometimes find it necessary to apply mineralogy to transfer the deposit from a mineral discovery to an ore deposit. The Lee Creek phosphate mine, which has been producing phosphate rock and phosphate fertilizers since 1966, was near abandonment but for the assistance of the North Carolina Research Laboratory under the direction of Imneo Redeker, plus the help of Paul Kerr. Paul Kerr and his student, Tom Rooney, discovered that deleterious calcium within the phosphate pellets were tied to the francolite lattice and could not be removed by ore dressing methods. Therefore, Texasgulf's program at Lee Creek was changed from a rock production plant to a calcined, high grade fertilizer complex. Kerr and Rooney further demonstrated that the best ore dressing technique could never bring the product to above 30.5% P_2O_5. Redeker at the Research Laboratory obtained this value in pilot plant and the 100 million dollar plant was based on Kerr's 30.5% P_2O_5.

Don Hausen also had the assistance of Paul Kerr in applied mineralogy at the Carlin Gold mine in Nevada. Although Hardie, McQuiston, and Hernlund described the deposit in 1965, Hausen and Kerr's paper in 1967 was a landmark dissertation on the nature of the fine gold occurrences which are now being discovered throughout Nevada.

Paul Kerr had a remarkable ability to be involved with major exploration programs within North America. His work on porphyry copper clay alteration was completed in 1952 and contributed to the great surge in porphyry exploration which commenced soon thereafter.

He worked with Jenney and myself in that unforgetable exploration boom in Canada, commencing in 1955. Many of us profited from his guidance in the fields of research and exploration for uranium. Paul Kerr, John Gruner, George Bain, and Lee Stokes were the first university professors involved in uranium research programs in 1951. Only Paul Kerr remained to the end of search in 1978. The current great American exploration effort is on fine volcanic gold in Nevada, of which he was one of the earliest investigators.

Paul Kerr had the uncanny ability to bring out the best in every student, most of whom developed into true scientists. However, a few of us were applied scientists, or possibly just explorer promoters, whom he attempted to nurture so that our civilization may have adequate minerals for the foreseeable future.

INNOVATIONS IN METHODOLOGY

Chapter 2

PROGRESS IN HIGH PRESSURE-TEMPERATURE MINERALOGY

by William A. Bassett

Department of Geological Sciences
Cornell University
Ithaca, New York 14853

ABSTRACT

Two very intense sources of electromagnetic radiation are contributing significantly to experimental studies of minerals at high pressures and temperatures: 1) A Q-switched YAG laser is able to produce a pulse of infrared radiation capable of heating a sample under pressures up to 400 kbar in a diamond cell up to temperatures in excess of $4000^{\circ}C$. 2) X-rays from the Cornell High Energy Synchrotron Source (CHESS) are used to obtain diffraction patterns of samples in a diamond cell in less than two minutes, patterns that would require hundreds of hours on a conventional source. The laser heating technique is being used to study the pressure-temperature conditions over which the melting of diamond takes place. In the initial stages of this investigation, diamond has been melted and techniques for detecting the melting have been developed. The intense X-rays from the synchrotron source have been used to examine the mechanism and rate of the pressure-induced phase transformation in Fe_2SiO_4.

INTRODUCTION

Diamond anvil cells have been in use since the early 1960's when A. VanValkenberg (1963) and his colleagues at the National Bureau of Standards made the first observations with such a device. Since that time, there have been a surprising number of analytical techniques from X-ray diffraction to electrical resistance used to study microscopic samples at high pressures and temperatures (Jayaraman, 1983). Many of these techniques have been applied to problems in geophysics (Bassett, 1979).

The very simplicity of the principle of the diamond cell allows a wide variety of means of gaining access to the sample while it is under pressure and at high temperature. The forces required to produce pressures of hundreds of kilobars, pressures comparable to those in the lower mantle of the Earth and even the core, are modest because of the small size of the sample. The diamond anvils, themselves, are ideal because they are transparent to light and to X-rays in addition to being the hardest material known. As a result, there are countless designs of diamond cells, each a little different in order to accommodate the particular analytical technique used.

In the diamond anvil cell, a sample is placed between flat, parallel, opposed faces on two perfect single crystal diamond anvils. Each anvil face is 0.3 to 0.6 mm in diameter. The anvils which are usually located on the ends of sliding pistons are driven together by means of a screw. In some experiments the sample may simply be placed between the anvils as a powder, in other experiments the sample may be mixed with a pressure medium such as NaCl, and in still other experiments the sample may be placed in a liquid held within a gasket between the anvil faces.

In the two experiments described in this paper, electromagnetic radiation, infrared in one case and X-rays in the other, was introduced through the diamond anvils. In the first experiment the sample was examined by electron microscopy after removal from the diamond cell. In the second experiment, X-ray diffraction patterns were obtained while the sample was being subjected to high pressures and temperatures.

MELTING OF DIAMOND

Carbon is a common element in the Solar System, and therefore relatively large quantities almost certainly exist within the Earth's interior. It has been suggested by Bundy (1980) that the melting temperature of diamond decreases with increasing pressure analogous to ice (Figure 1). Dickey et al. (1983) discuss the possibility that this effect may result in the existence of a carbon melt within the Earth's lower mantle. The purpose of the project described in this paper is to test this hypothesis. This research has been carried out at Cornell University in collaboration with Jon Gold, John M. Bird, and Maura S. Weathers (Gold et al., 1984). The first stage of this project has consisted of assembling a laser powerful enough to melt diamond. The second stage has been to establish suitable criteria for determining if melting took place. The third stage will consist of constructing apparatus to measure the spectrum of the incandescent light from the melting diamond so that temperature can be calculated.

The laser heating apparatus is shown in Figure 2. It consists of

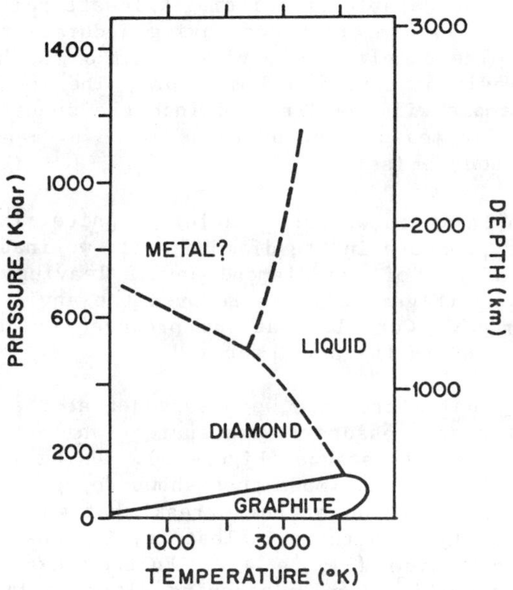

Fig. 1. Carbon phase diagram. The dashed lines are postulated.

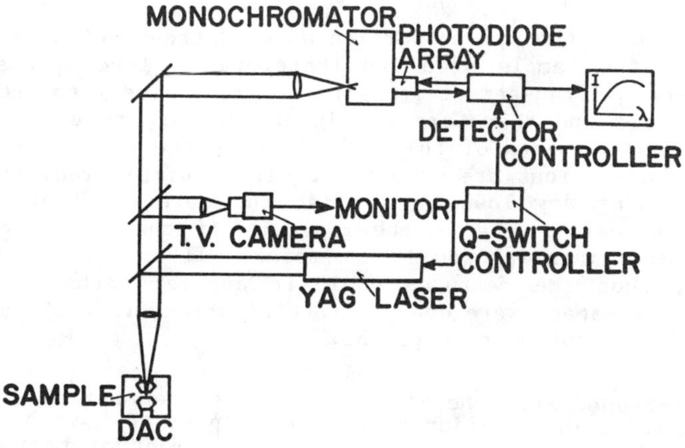

Fig. 2. Apparatus for laser heating and temperature measurement.

a Q-switched YAG laser capable of emitting kilowatt pulses of infrared with a wavelength of 1.06 microns and having a duration of a few hundred nanoseconds. The sample can be viewed during the heating process by means of a television camera and monitor. The incandescent light from the heated sample will be directed into the spectrometer and the spectrum recorded for measurement of temperature by means of the sample´s quasi black body emission.

During a run in which we were heating graphite mixed with potassium bromide under pressure in the diamond cell, we inadvertantly damaged the surface of one of the diamond anvils leaving a streak about 0.1 millimeters long (Figure 3). We removed the anvil to examine the nature of the damage. Our observations produced evidence of direct melting of the diamond in the anvil face.

Under the optical microscope we observed that the portion of the streak that was at high pressure is transparent while the portion that was at a lower pressure is opaque (Figure 3). Using a scanning electron microscope (SEM) set in a mode that shows topography we were able to determine that the portion of the streak that was at higher pressure has the topography of a furrow. That is, it consists of a trough with ridges on each side (Figure 4). Further examination of the trough at higher magnification in scanning electron image (SEI) mode revealed numerous bright spots which when analyzed were found to be potassium bromide (Figure 5). The surface of the diamond anvil was then washed in water to dissolve the potassium bromide. When the surface was examined again by scanning electon microscope it was found that some of the bright spots remained. On closer examination, these were found to be spheres of potassium bromide imbedded in the carbon. A stereo pair of images taken in SEI mode confirmed that they lie beneath the surface of the carbon.

We interpret the opaque portion of the streak which was closer to the edge of the sample area and therefore at lower pressure to be graphite and the transparent portion that lies closer to the center of the sample area and therefore at higher pressure to be diamond. Although exact meaurments of the topography of the furrow are not feasible, our observations are consistent with a simple redistribution of material without any loss. That is, the material that was in the trough now appears to be in the ridges. If the damage were due to oxidation, some material would be missing. If the damage were due to fracturing, then the features would be angular rather than smooth. Finally if the damage were due to graphitization, then the material of the furrow would not be transparent.

The imbedded KBr spherules are perhaps the most persuasive evidence of melting of the diamond. These appear to have been droplets of the KBr that were entrapped during the moment of melting of both the diamond and the KBr.

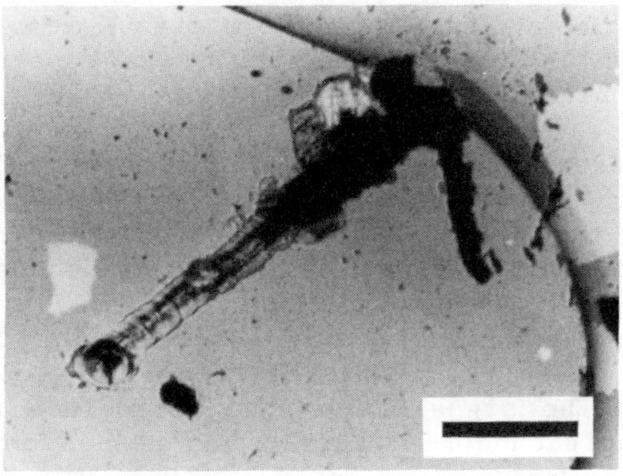

Fig. 3. Transmitted light micrograph of the damage streak on the surface of one of the diamond anvils produced by radiation from a YAG laser focused to a 10 micron spot. The opaque portion was at lower pressure than the transparent portion. (scale bar = 40 microns)

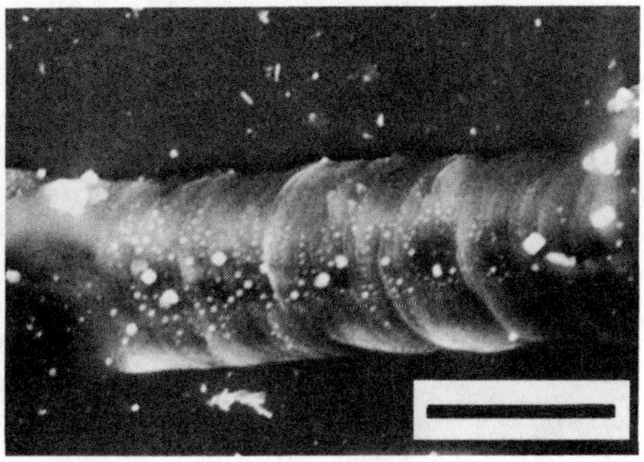

Fig. 4. (Left) Scanning electron image of the streak using a back-scattered electron image mode designed to enhance the topography. Note the furrow-like feature at the higher pressure end (upper left). (scale bar = 30 microns)

Fig. 5. (Right) Secondary electron image of the furrow. The bright spots in the furrow are spherules of KBr. Many of the spherules remained after washing with water indicating that they are completely encapsulated in the carbon. (scale bar = 12 microns).

The experiment was repeated but this time with only KBr between the diamond anvils. A similar damage streak resulted. A SEM image revealed a trough, the bottom of which was covered with nodules of carbon. We interpret these to be quenched droplets of molten diamond.

With the experiments described above, we believe that we have established reliable criteria for recognizing when melting has taken place. Our next task will be to measure the temperature during melting. To accomplish this we will use the grating spectrometer shown in the upper part of Figure 2 to spectroscopically analyze the incandescent light from the sample. The spectrum will be picked up by a photodiode array which is capable of recording a complete spectrum in 20 nanoseconds. The signals from the photodiode array will then be collected in a multichannel analyzer and stored on a floppy disk for future analysis. Preliminary tests of this equipment have convinced us of the feasibility of obtaining reasonably accurate temperatures by this method.

MECHANISM OF THE OLIVINE-SPINEL TRANSITION

There are currently two theories concerning the nature of the transition from the olivine structure (α) to the spinel structure (γ) at high pressure and temperature in orthosilicates. Sung and Burns (1976) have proposed that the transition takes place by incoherent nucleation and recrystallization. Poirier (1981), on the other hand, has suggested that the transition takes place by a martensitic type of mechanism in which a restacking of the oxygen atoms from a hexagonal close packed arrangement to a cubic close packed arrangement is accompanied by a reordering of the cations. In subsequent studies some investigators have found evidence in support of incoherent nucleation and recrystallization (Nakajima and Hamaya, 1982; Vaughan et al., 1982; Boland and Liebermann, 1983). Other investigators have found evidence in support of the martensitic-like mechanism (Lacam et al., 1980; Hamaya and Akimoto, 1982; Shimizu et al., 1982; Boland and Liu, 1983). The project described next in this paper was undertaken in collaboration with Michael D. Furnish in an attempt to resolve this controversy (Furnish and Bassett, 1983). The very intense white X-rays available at the Cornell High Energy Sychrotron Source (CHESS) was utilized to make diffraction patterns of Fe_2SiO_4 while the phase transition from the olivine structure to the spinel structure was in progress.

A diamond cell was designed with an internal resistance heater wound around the metal supports for the diamond anvils. A thermocouple placed in contact with one of the anvils was used to measure the temperature. A pin hole with a diameter of approximately 150 micron was used to collimate the X-ray beam so that it would pass through the center of the sample while the sample was being subjected to high pressure and temperature. The scattered X-rays were picked up by an intrinsic germanium detector and analyzed by an energy dispersive technique. The data were then collected in a multichannel analyzer and stored on a floppy disk.

Because of the intensity of the radiation from the synchrotron source, the experiment had to be placed in a well shielded radiation cave. Both pressure and temperature were controlled remotely. The pressure was increased by a selsyn motor which applied torque through a series of gears to a screw which in turn applied a force on the diamond anvils. The temperature was controlled by remote transformers. This capability allowed the operator to increase pressure and temperature without having to enter the cave. As a result, a complete cycle including recording a diffraction pattern and advancing the pressure and/or temperature could be accomplished in 2 minutes.

In Poirier's (1981) model for the phase transition in olivine he proposed that the restacking of the oxygen atoms would be simultaneously accompanied by the reordering of the cations. We suggested, however, that there might be a delay between the two processes that would be detectable if diffraction patterns could be collected fast enough. During that delay, the oxygen atoms would be in a cubic close packed arrangement but the cations would still remain in their original sites or be "in transit" to their new sites. Cubic close packed oxygens have a repeat distance of approximately 4A, whereas the spinel structure in which the cations are ordered has a repeat distance of approximately 8A. If there is an intermediate stage in which the oxygens are in a cubic close packed arrangement but the cations are not ordered, then only those reflections that have indices h/2, k/2, and l/2 all even or all odd should be present. It was this effect that we looked for and found (Furnish and Bassett, 1983). Figure 6 shows a series of diffraction patterns progressing from bottom to top that were taken as pressure was increased across the transition at 520°C. Note that the 400 peak (which meets the criterion given above) appears first in the fourth pattern but the 331 peak (which does not meet the criterion) appears later in the fifth pattern. This effect was observed in two additional series. In a fourth series made at 400°C we turned off the heat and quenched the sample as soon as the 400 peak appeared and then transferred the sample to a Debye-Scherrer camera for greater resolution. The resulting pattern contained the 400 and other peaks that meet the criterion but lacked the 331 and other peaks that do not meet the criterion.

We have interpreted our observations as indicating that the olivine-spinel phase transition takes place by means of a martensitic-like restacking mechanism as suggested by Poirier (1981). The stress in a diamond cell is very anisotropic. Sample extrudes from between the diamond anvils as pressure is applied and in the process is subjected to a great deal of shear stress. In other high pressure devices, the stress is more isotropic. This difference in stress environments may account for the diverse findings of experimentalists

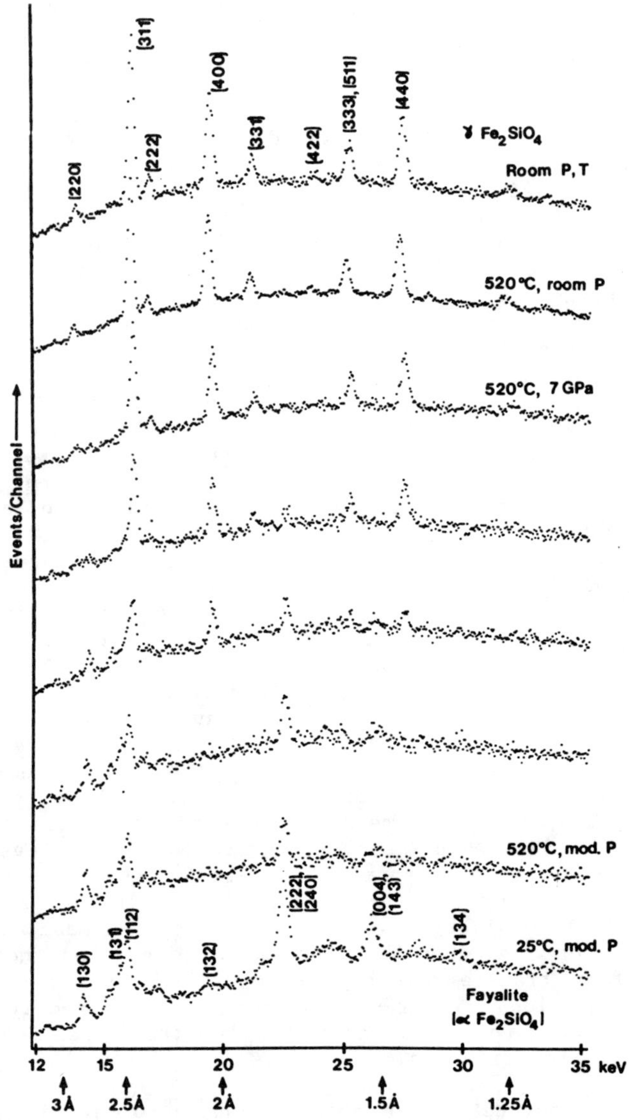

Fig. 6. X-ray diffraction spectra selected from a series acquired across the olivine(α)-spinel(γ) transition in Fe_2SiO_4 at 520°C. Note that the 400 peak of the spinel phase appears earlier (fourth spectrum from the bottom) than the 331 peak (which appears in the fifth spectrum from the bottom). The lapsed time between the two spectra is approximately 2 minutes.

who have conducted studies on this problem. We intend to conduct additional experiments to determine the extent to which the mechanism depends on the anisotropy of the stress.

REFERENCES

Bassett, W. A., 1979, "The Diamond Cell and the Nature of the Earth's Mantle," Annual Review of Earth and Planetary Sciences, Vol. 7, pp. 357-384.

Boland, J. M. and Liu, L. G. 1983, "Olivine to Spinel Transformation in Mg_2SiO_4 via Faulting Structures," Nature, Vol. 303, pp. 233-235.

Boland, J. M. and Liebermann, R. C., 1983, "Mehcanism of the Olivine to Spinel Phase Transformation in Mg_2SiO_4," Geophysical Research Letters, Vol. 10, pp. 87-90.

Bundy, F. P., 1980, "The P,T Phase and Reaction Diagram for Elemental Carbon, 1979," Journal of Geophysical Research, Vol. 85, pp. 6930-6936.

Dickey, J. S., Bassett, W. A., Bird, J. M., and Weathers, M. S., 1983, "Liquid Carbon in the Lower Mantle?" Geology, Vol. 11, pp. 219-220.

Furnish, M. D. and Bassett, W. A., 1983, "Investigation of the Mechanism of the Olivine-Spinel Transition in Fayalite by Synchrotron Radiation," Journal of Geophysical Research, Vol. 88, pp. 10333-10341.

Gold, J. S., Bassett, W. A., Weathers, M. S., and Bird, J. M., 1984, "Melting of Diamond," Science, Vol. 225, pp. 921-925

Hamaya, N. and Akimoto, S., 1982, "On the Mechanism of the Olivine-Spinel Phase Transition," Physics of Earth and Planetary Interiors, Vol. 29, pp. 6-11.

Jayaraman, A., 1983, "Diamond Anvil Cell and High Pressure Physical Investigations," Reviews of Modern Physics, Vol. 55, pp. 263-266.

Lacam, A. M., Madon, M., and Poirier, J. P., 1980, "Olivine Glass and Spinel Formed in a Laser Heated, Diamond Anvil High Pressure Cell," Nature, Vol. 288, pp. 155-157.

Nakajima, Y. and Hamaya, N., 1982, "Olivine-Spinel Transformation Mechanism in Single-Crystal Ni_2SiO_4," (Abstract) EOS Transactions of the American Geophysical Union, Vol. 63, p. 431.

Poirier, J. P., 1981, "Martensitic Olivine-Spinel Transformation and Plasticity of the Mantle Transition Zone," Anelasticity in the Earth,

Geodynamics Series, Vol. 4, F. D. Stacey, ed., American Geophysical Union, Washington, D.C., pp. 113-117.

Shimizu, H., Bassett, W. A., and Brody, E. M., 1982, "Brillouin-Scattering Measurements of Single-Crystal Forsterite to 40 kb at Room Temperature," Journal of Applied Physics, Vol. 53, pp. 620-626.

Sung, C. M. and Burns, R. G., 1976, "Kinetics of High Pressure Phase Transformations: Implications to the Evolution of the Olivine-Spinel Transition in the Downgoing Lithosphere and its Consequences on the Dynamics of the Mantle," Tectonophysics, Vol. 31, pp. 1-32.

VanValkenberg, A., 1963, "High Pressure Microscopy," High Pressure Measurement, A. Giardini and E. C. Lloyd, ed., Butterworth, Washington, pp. 87-94.

Vaughan, P. J., Green, H. W. (II), and Coe, R. S., 1982, "Is the Olivine-Spinel Transformation Martensitic?," Nature, Vol. 298, pp. 357-358.

ACKNOWLEDGMENTS

The author wishes to thank his colleagues, John M. Bird, Maura S. Weathers, Michael D. Furnish, Eugene Huang, and Jon S. Gold for their contributions to the research reported in this paper. The work was supported by National Science Foundation grants EAR8313569 and EAR81-16083.

Chapter 3

TEM MEASUREMENT OF ORGANIC SULFUR IN COAL MACERALS

K.C. Hsieh and C.A. Wert

Materials Research Laboratory
and
Department of Metallurgy and Mining Engineering
University of Illinois, Urbana, IL 61801

ABSTRACT

Measurement has been made of the organic sulfur concentration in coal macerals using microanalytical techniques in the transmission electron microscope. This technique utilizes measurements of x-rays emitted by the sulfur atoms embedded in the hydrocarbon constituents and the bremsstrahlung x-rays generated in the matrix. It is a simple technique for thin specimens and yields reliable results. Although it can be used to measure the average organic sulfur content of coal, it is most useful in determining organic sulfur in extremely small volumes of material. Variations in organic sulfur content can be measured at a scale unimagined earlier. This TEM technique also has application for determination of organic chlorine, phosphorous, potassium and other elements often found distributed through the hydrocarbon constituent of fuels.

INTRODUCTION

Direct measurement of the organic sulfur content of coal macerals can be made by application of several electron-optical methods. Use of the electron microprobe has been demonstrated by several investigators, see for example Solomon and Manzione (1977), but the most prominent group has been that associated with Raymond et al. (1978, 1979, 1982). The technique developed by these investigators avoids the separate measurements of total sulfur and mineral sulfur required by the standard ASTM technique. Even so, some difficulties attend application of the electron microprobe because the area of examination is rather large, caused by electron spreading in a thick sample. We have demonstrated that ultra-thin specimens such as those used in the transmission electron microscope avoid some of these problems and provide means of analysis for organic sulfur in a volume less than 1 μm^3.

The method has been adapted from earlier biological work pioneered by Hall and his collaborators (1971, 1973, 1979). The presence of the element being measured, in our case S, is determined using the Kα x-ray emission line. The count rate for that line, C_s, is made using the output of a standard energy-dispersive detector. This count rate is proportional to the concentration of S in the volume being irradiated by the electrons. To determine that volume, one needs an independent measurement. Hall showed that this could be a simultaneous determination of the count rate, C_b, of the bremsstrahlung radiation emitted from that same volume. For a thin foil, the concentration of organic sulfur is proportional to the ratio of these two count rates, C_s/C_b. Using appropriate standards one can calibrate the system to give weight fraction sulfur. Details of this investigation are described in the next section.

EXPERIMENTAL METHOD

The measurement itself is simple. A thin foil or powdered specimen is inserted into the microscope in a standard specimen holder. The geometry is shown in Fig. 1. The presence of organic sulfur in the hydrocarbon material is determined by noting the presence of the sulfur Kα-line using the multichannel analyzer associated with the detector. Sulfur in mineral form can be easily avoided since the electron beam can be placed on a spot far removed from pyrite or other minerals. Such mineral forms show up sharply dark in the field of view of the electron microscope. Thus, one can be certain that sulfur in pyrite is not being counted.

Simultaneously with the sulfur count rate, the bremsstrahlung count rate is measured over some energy interval. An appropriate energy interval lies between 10 and 18keV, since few characteristic lines of impurities found in coal exist over that range. These two count rates are sketched in Fig. 2. The organic sulfur content is proportional to the ratio of two shaded areas. Several features are required to verify that this system works. First, the ratio of C_s to C_b should be independent of the foil thickness, and this has been demonstrated by the measurements in Fig. 3. The ratios for films of pure sulfur, dibenzothiophene, and dibenzothiophenesulfone are shown here. The lines in Fig. 3 are straight for the thin foils used in this analysis, and thus, the measurements are confirmed to be independent of foil thickness. However, the slopes differ, because the sulfur content of the three materials varies. The slopes of these three lines are consistent with the relative sulfur concentration in the three materials. Thus each can be used to provide an independent calibration factor. The average of the three factors results in the following equation:

$$S\ (\text{wt }\%) = 1.6\ C_s/C_b$$

This linearity holds only for thin films and the number 1.6 is applicable to measurements for bituminous coals. Other microscopes

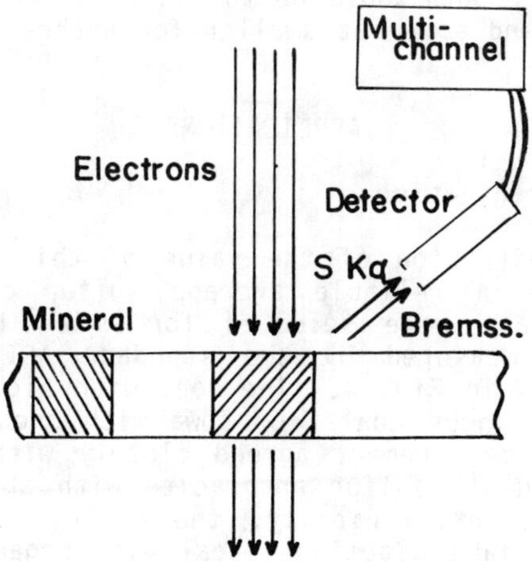

Figure 1. Geometry of X-ray measurement.

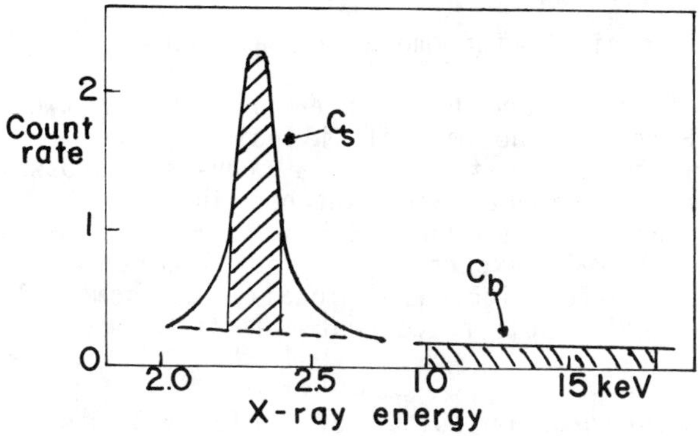

Figure 2. Sketch of count rates C_s and C_b.

with different geometries might have constants slightly different from 1.6. This constant would be somewhat larger for sub-bituminous and lignite coals and somewhat smaller for anthracites, but only by a few percent.

APPLICATIONS

Average Sulfur Concentration

The first demonstration of the value of this method showed that the technique gave a reliable average sulfur content. For this purpose, three coals were selected for which the organic sulfur content had been measured by the standard ASTM technique. The comparison is shown in Fig. 4. The coal with highest organic sulfur content was a bituminous coal from Iowa with organic sulfur of more than 4%. The TEM measurement agreed closely with that of the ASTM method. A low organic sulfur anthracite with about 1% sulfur also showed close agreement. Finally, the organic sulfur content was measured for an Indiana bituminous coal with organic sulfur at about 0.6%. Again the agreement was excellent. In conclusion, these measurements, averaged over the 50 independent observations made over different regions of each of these coals, gave excellent quantitative analysis for organic sulfur.

The insets in Fig. 4 showed that the spread of observations about the means was surprisingly large. Since this spread is not experimental error, it must reflect real variations in organic sulfur content among different volumes of the material. Accordingly, a number of measurements were made to examine this effect in more detail.

Variation of of Organic Sulfur Among Maceral Types

The origin of this spread has been deduced by two independent measurements. First, a specimen of Illinois #5 coal was thinned and a series of measurements was made on a traverse across a vitrinite maceral. Those measurements are sketched in Fig. 5. The average organic sulfur content was about 2.5 wt%, with a small variation about the mean. A small excursion was made laterally about Point 29 and the organic sulfur increased locally to somewhat above 3%. Overall, the organic sulfur content of this maceral was quite uniform.

The variation of organic sulfur of whole coal was also measured among maceral fractions of separated coal. The fractions were separated by a centrifuge-gradient method used at the Argonne National Laboratory, Dyrkacz et al. (1984). Four coals were examined and their properties are listed in Table 1. The coals were finely ground and separated by the density gradient method to give density fractions between 1.0 and 1.5 grams per cc. The maceral types of each of these coals were determined by standard optical reflectance

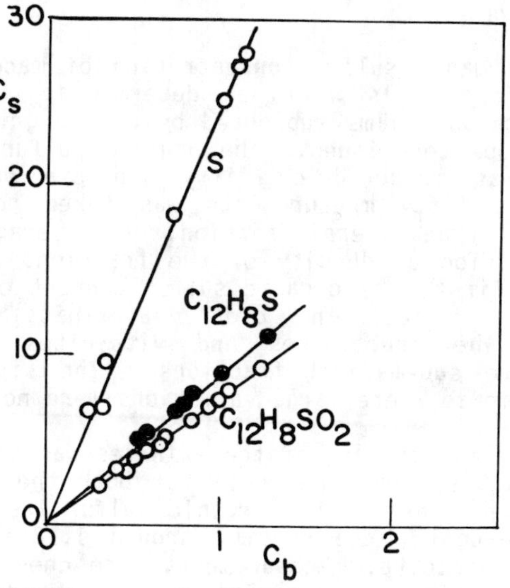

Figure 3. Calibration measurements.

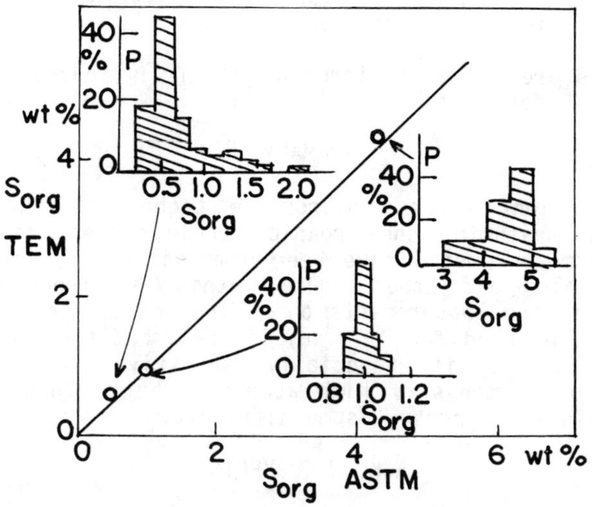

Figure 4. Comparison of S_{org} measured by TEM techniques and the ASTM standard method.

techniques, so that the maceral-type associated with each density fraction was known.

The average organic sulfur concentration of maceral fractions of each of these four coals was then determined. The powders were placed on thin carbon films supported by copper grids using standard electron microscope techniques. The organic sulfur concentration of each fraction was measured on fifty separate particles and the average of these fifty measurements was taken to be the organic sulfur content of that maceral fraction. The averages are plotted in Fig. 6 as a function of density of the fraction. Several features were observed. First, the organic sulfur content of the three major maceral groups varied in order S(exinite)>S (vitrinite)>S (inertinite). The inertinites and vitrinites showed not much variation for the sub-maceral fractions. The situation was quite different for exinite where large variations were noted.

The organic sulfur content of the exinites was sharply peaked at a density of about 1.18 grams per cc. For both the resinites and the alginites of lower density, the organic sulfur was much lower. For these bituminous coals, the density around 1.18 was close to that associated with sporinite. Measurements were checked on an Illinois #5 coal which contained a sporinite maceral embedded in vitrinite, Fig. 7. The sporinite had a much higher organic sulfur content than the surrounding vitrinite macerals. Furthermore, the value of 4.5 wt% in Fig. 7 was about the value at the peak of Illinois #5 in Fig. 6, confirming the assumption that the density fraction near 1.18 was sporinite. Also, the average value for the vitrinite, near 2.5, was the same as that for the vitrinites in Fig. 6.

Measurements are being confirmed on resinites and fusinite in this same Illinois #5 coal.

SUMMARY

The transmission electron microscope technique shows remarkable capability for measuring the organic sulfur content of hydrocarbon fuels. These measurements have been demonstrated for coal, as well as for oil shales and amber. The method is direct and has the capability of making measurements on a fine scale. In principle, the technique could be used for S in any mineral, but the proportionality constant in Eq. 1 is especially sensitive to the relative concentration of elements of high atomic number. Thus, sulfides of the heavy metals would present some difficulty.

ACKNOWLEDGMENTS

Support for this work came from the Division of Materials Research, DOE, under contract DE-AC02-76ER01198. Part of the support for K.C. Hsieh also came from the Center for Research on Sulfur in Coal affiliated with the University of Illinois. The separated coals

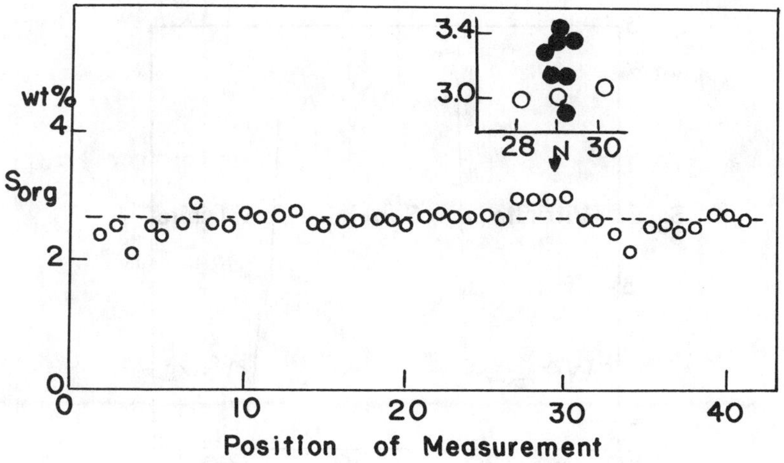

Figure 5. S_{org} in a traverse of a vitrinite maceral in a specimen of Illinois #5.

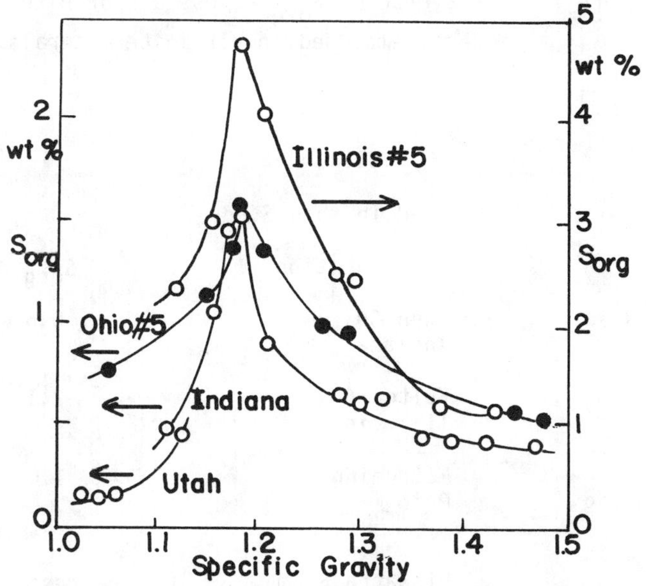

Figure 6. S_{org} as a function of specific gravity of maceral fractions.

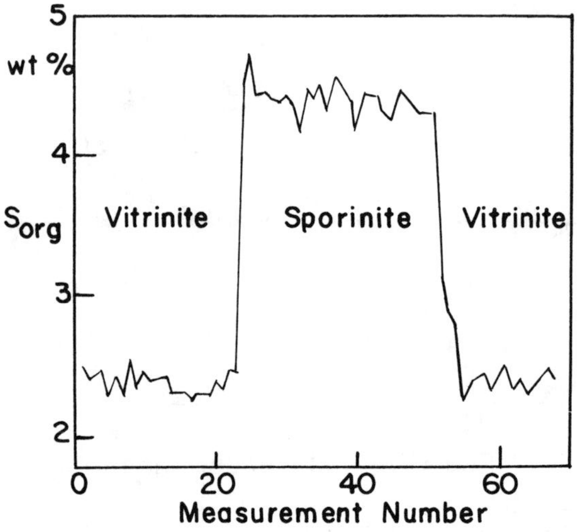

Figure 7. A traverse of S_{org} across a sporinite maceral embedded in vitrinite macerals.

Table 1

Coals Used in this Study

Coal	Location	S_{org} daf
Indiana Block #1 PSOC 106	Owen Co. Indiana	0.6 wt.%
Illinois #5 PSOC 592	Fulton Co. Illinois	2.1
Ohio #5 Lowert Kitanning PSOC 297	MaCroning Co. Ohio	1.0
Bituminous SIU-561	Hiawatha Seam Utah	resinite
Illinois #5	Galatin Co. Illinois	2.15

were supplied by Gary Dyrkacz. The data in Fig. 6 were made by M. Buckentin for his MS thesis; a complete discussion of his results will be published later.

REFERENCES

Dyrkacz, G.R., Bloomquist, C.A.A. and Ruscic, L., 1984, Fuel, 63, 1166.

Hall, T.A., 1971, Physical Techniques in Biological Research: Optical Techniques, 2nd Ed., Vol. 1A, G. Oster, ed., 157.

Hall, T.A., 1979, J. of Microscopy, 117, Pt. 1, 145.

Hall, T.A., Anderson, H.C. and Appleton, T., 1973, J. of Microscopy, 99, Pt. 2, 177.

Raymond, R., Jr. and Gooley, R., Scanning Electron Microscopy/1978/I, 93.

Raymond, R., Jr., 1979, Proc. Annual Conf. Microbeam Analytical Soc., 14, 93.

Raymond, R., Jr., 1982, ACS Symposium Series, No. 204, COAL AND COAL PRODUCTS; ANALYTICAL CHARACTERIZATION TECHNIQUES. E.L. Fuller, ed., Amer. Chem. Soc., Washington, D.C.

Solomon, P.R. and Manzione, A.V., 1977, Fuel, 56, 393.

Chapter 4

ORE.ID, A MICROCOMPUTER PROGRAM FOR THE RAPID IDENTIFICATION OF OPAQUE MINERALS IN ORES AND BENEFICIATION PRODUCTS

by John E. Hagni and Richard D. Hagni

Computer Analyst, Delta Research Corporation
Arlington, Virginia

Gulf Oil Foundation Professor of Geology, University of Missouri-Rolla
Rolla, Missouri

ABSTRACT

ORE.ID, a microcomputer program, rapidly searches a database of the principal properties for 573 ore minerals. Searches may be performed on any or all of the following properties: mean Vickers hardness, Talmage hardness, mean reflectance, strength of anisotropism, and colors and tints of the mineral and its internal reflections. Data from the International Mineralogical Association - Commission on Ore Microscopy data card file have been used when available. The program displays the number and type of data matches for each mineral found, displays their chemical components, and allows the range to be narrowed or widened for each of the identification properties for repeated searches, if desired. ORE.ID is much more rapid, allows for the use of more identification properties, and involves more ore minerals than any previous identification system.

INTRODUCTION

The past seven years or more has witnessed a striking improvement in amount of quantitative data available for ore minerals. Measured reflectance values are now available for 432 ore minerals. Vickers hardness numbers have been measured for 316 ore minerals. Quantitative color coordinates have been determined for 107 ore minerals. Although quantitative measurements of the rotation properties, apparent angle of rotation and the ellipticity, are available for 125 ore minerals, their use has fallen into disfavor.

Not only have the numbers of data increased but the quality of those data have become much more reliable through the increased use of internationally accepted reflectance standards and procedures. Such standardized data have now been published for 159 ore minerals, available in the International Mineralogical Association - Commission on Ore Microscopy (IMA-COM) quantitative data cards (Henry, 1977).

Now that sufficient quantities of high-quality quantitative data are available, there is an urgent need to devise an ore mineral identification scheme that utilizes these data, and information on the qualitative properties of those minerals, to provide a practical, workable, comprehensive, and accurate means of ore mineral identification. Seven years has passed since the publication of the IMA-COM data cards and no satisfactory system for ore mineral identification utilizing those data has been forthcoming.

The purpose of this paper is to discuss a microcomputer program that has been written to utilize the quantitative data in the IMA-COM data cards and other quantitative and qualitative data available for ore minerals in the literature, to rapidly search for minerals whose properties are similar to that of an unknown under investigation. The program is easy to use, contains nine types of information on 573 minerals, completes the average search in fifty seconds to two minutes, and indicates the number and character of data matches for each of the minerals found by the computer. The program has been designed to narrow the number of possible minerals whose properties fit a given unknown so that those minerals can be considered further in standard descriptive volumes on ore minerals and in the published literature. For many ore minerals, however, their identity is evident at the conclusion of the computer search.

PREVIOUS IDENTIFICATION SYSTEMS

Ore microscopists have long strived for an adequate system for the identification of ore minerals in polished sections. All of the previous systems have suffered from one of more of the following problems: 1) arrangements that are systematic or alphabetical, 2) the lack of sufficiently standardized quantitative data, 3) the inclusion of an insufficient number of ore minerals, 4) the requirement of the measurement of certain properties that are not routinely measured by most ore microscopists, 5) the utilization of quantitative data to the exclusion of the equally useful qualitative information on ore minerals, and 6) the tedious and slow manner in which an ore mineral is identified.

Some of the older systems of ore mineral identification depended upon the determinations of one or two largely qualitative properties, such as their color and scratch hardness (Murdoch (1916), and the colors of mineral, powder, and internal reflections (Van der Veen, 1925). The early systems of Davy and Farnham (1920), Farnham (1931), and Short (1940) were based principally upon the reaction of ore minerals to drops of acid placed upon the surface of the polished section. Systems that relied mostly upon qualitative color neglect other important diagnostic properties of ore minerals. Those systems that relied almost entirely upon etch reactions have fallen into disuse due to the inconsistent results derived from the reactions of certain ore minerals with the selected acids, the advent of quality reflectance and indentation hardness measurements, and the fact that etched sections required repolishing.

The punched card system of Fairbanks (1946) also utilized the results from reactions with acids, but also included ore mineral data on color, reflectance, and megascopically determined properties on 117 cards.

The system of Chamot and Mason (1940), which utilizes microscopic examinations of the optical properties of the products of qualitative chemical reactions, also has fallen into disuse due the requirement of gouging a powder out of the surface of the polished section with a needle and the advent of electron microprobes and scanning electron microscopes with energy dispersive spectroscopic analysis systems (SEM-EDS).

More recent ore microscopic identification systems have employed a greater degree of quantitative measurements. Schneiderhöhn (1952), for example, contains a determinative table that is based mainly upon quantitative reflectance and partly upon qualitative hardness, color, and anisotropism. Cameron (1961) has provided lists of some of the more common ore minerals arranged by reflectance, Vickers hardness, and rotation properties, as well as by polishing hardness and the color of the internal reflections. Craig and Vaughan (1981) have furnished lists arranged by reflectance, Vickers hardness, and color. Some of the most useful quantitative systems are those charts and tables that are based upon Vickers hardness and reflectance (Bowie and Taylor, 1958; McLeod and Chamberlain, 1969; Tarkian, 1974; Galopin and Henry).

The most intensively used volume for the study of ore minerals is that of Ramdohr (1980). That reference contains the most useful descriptions of ore minerals, but the minerals are arranged according to the Dana system, rather than by increasing Vickers hardness or reflectance (although a table of reflectance values from three sources for many minerals is contained in a pocket in Ramdohr). The common practice has been to narrow the possible minerals for a given unknown down to three or four by some other system, such as that of McLeod and Chamberlain (1969) or Uytenbogaardt and Burke (1971) and then consult Ramdohr's volume for the detailed descriptions of those minerals.

Similarly, the two volumes by Picot and Johan (1977;1982) form excellent descriptive volumes for many ore minerals, but the minerals are arranged alphabetically. The latter volume contains descriptions for 369 ore minerals, provides reflectance values at every 20 nm between 420 and 700 nm for 334 ore minerals, and makes special use of photomicrographs showing the colors of anisotropism. Although those authors provide determinative tables to accompany the descriptions, the tables rely greatly upon reflectance and neglect Vickers hardness and many other useful properties for the identification of ore minerals.

The system of Uytenbogaardt and Burke (1971) has been one of the most successful schemes for ore mineral identification. It contains a sufficiently large number of ore minerals and its arrangement is largely on the basis of Vickers hardness together with ore mineral association.

Eales (1969) has proposed a scheme based upon the measurement of the rotation properties, apparent angle of rotation and the ellipticity. Data for 38 isotropic and uniaxial minerals were presented in chart and tabular form.

Most importantly, all of the above volumes either have been published prior to that of the IMA-COM quantitative data file in 1977 or did not utilize data from that source. There have been four or five ore mineral identification systems that have been proposed since 1977. The first was that of Bowie and Simpson (1978) who utilize reflectance values at four standard wave lengths, which are shown on four colored charts. The Vickers hardness numbers for those ore minerals are presented only in a separate table. Because the reference is restricted to data for only 37 ore minerals, its use is largely limited to introductory college laboratories.

An index or set of five determinative tables was prepared by Leonard (1979) to be used in conjunction with the IMA-COM data file. His scheme involved the use of the quantitative bireflectance, the positive or negative slope of the spectral reflectance curve, and the reflectance at 546 nm. Although this scheme is an innovative one, the first two properties are ones that many ore microscopists do not measure on a routine basis.

A punched card system, the Delft System, by Kühnel, Prins, and Roorda (1980) provides both quantitative and qualitative information on about 195 ore minerals. Reflectance data is provided in an alphabetical list for 217 additional minerals. Use of the system is cumbersome and slow, and the volume does not include the IMA-COM data.

The only ore mineral identification systems to use a computer were those of Bernhardt (1979) and Atkin and Harvey (1979a, 1979b, 1982). The system of Bernhardt involves the comparison of the spectral reflectance curve determined for an unknown mineral with the curves for a large number of ore minerals that have been previously determined and stored in a computer program. The program of Atkin and Harvey operates on a Radio Shack TRS-80 computer and utilizes quantitative color coordinates and Vickers hardness number to search a database of 240 ore minerals. The system requires that a total of 16 measurements of reflectance be taken at 20 nm intervals, and that the quantitative color coordinates, x, y, Y, Pe, and λ_d, be calculated from those measurements. Alternatively one can purchase a motorized diffraction grating monochrometer to interface with the microcomputer. For the first alternative, a significant amount of time is consumed for both the measurements and the calculations; for the second alternative, an expense is involved that may be beyond that which some laboratories may wish to expend for this purpose.

As part of a short course workbook on Applied Ore Microscopy, Hagni (1984) has prepared three tables for the identification of ore minerals. One table is arranged alphabetically, one by increasing Vickers hardness, and the third by reflectance. The tables also include the available quantitative data on color coordinates, and the rotation properties, angle of rotation, ellipticity, and sign of retardation, and qualitative information on color, letter hardness,* bireflectance, anisotropism, colors of anisotropism, and intensity and colors of the internal reflections. Each of the tables contains the available data on 573 ore minerals.

At the time that most of the above systems were developed, microcomputer technology was not in widespread use. In fact, during the time that the two previous computer systems were being developed, the use of relational database management systems on microcomputers was virtually non-existent. The microcomputer program discussed in this paper utilizes such a system to search a database consisting of both quantitative and qualitative information for a large number of ore minerals.

THE MICROCOMPUTER PROGRAM

The microcomputer program, ORE.ID, involves the search of a database that consists of all the available information on nine properties (including chemical constituents) for 573 ore minerals. The data in the program have been taken from tables in Hagni (1984). Those tables summarize the available quantitative, semi-quantitative, and qualitative data for 573 ore minerals.

Quantitative data in the database include the mean values for Vickers hardness and reflectance. Qualitative or semi-quantitative information in the database include letter hardness,* the color and

---

*The term "letter hardness" is utilized in this paper to include: 1) Talmage hardness, the resistance of an ore mineral to scratch, and 2) polishing hardness, the resistance of an ore mineral to polish that is observed in the form of relief in polished sections, and 3) a hardness determined by comparison to Vickers hardness, the resistance of an ore mineral to indentation, and to Mohs hardness, the ability of one mineral to scratch another. Thus, the letter hardness used here has a close but not always strictly equivalent relationship with Talmage hardness.

tint of the ore minerals, the strength of anisotropism, and the primary color and tint of the internal reflections. Although the program does not search on the basis of chemical compositions, the chemical components for each of the minerals found are displayed to form a ninth criterion for identifying the unknown mineral.

ORE.ID is run on an IBM PC at the University of Missouri-Rolla, but it will operate on other microcomputers.

Vickers Hardness Number

The database contains the mean Vickers hardness number (VHN) for all of those ore minerals for which the information is available in the literature. For those ore minerals whose Talmage hardness is known but whose VHN has not yet been measured, a reasonable approximation of VHN has been entered into the database. For some ore minerals, the hardness has been measured in the University of Missouri-Rolla (UMR) ore microscopy laboratory.

The range of Vickers hardness that the computer will search for an unknown ore mineral has been pre-selected. The range depends upon the magnitude determined for the VHN of the unknown mineral. The VHN search ranges are:

±250	1000 - 3000
± 50	300 - 999
± 30	100 - 299
± 30	1 - 99

For example, for an unknown whose VHN has been measured to be 500 the computer will search for minerals with mean Vickers hardness numbers between 450 and 550. Those minerals will be selected that have Vickers hardness numbers within that range and that have other properties that also match those of the unknown.

The program has the option to change the above ranges of search. If an unsatisfactory number of possible minerals is found by the initial computer search, the operator may either narrow or widen the range of Vickers hardness number that is evaluated for repeated searches. If, for example, the computer finds too many possible minerals to fit the unknown, the user may wish to limit subsequent searches to a narrower range of VHN, particularly if he feels that his measurements of VHN are of superior quality. On the other hand, if the initial search finds no mineral that has a significant number of data matches, either the measurement of VHN for the unknown is not sufficiently accurate or the quality of the data in the literature is poor (or the mineral is new), and the search range for VHN can be widened for a repeated search of the database.

If the user has not determined Vickers hardness for the unknown ore mineral or he does not wish the computer to consider VHN for the database search, he can enter "no data" for this category.

Letter Hardness

Letter hardness data have been included in the database to allow those individuals whose laboratories lack the equipment for the measurement of Vickers hardness number to also have the opportunity to use the microcomputer program. Letter hardness can be readily determined in any laboratory under the microscope by: 1) utilizing the width of the ever-present scratches, 2) observing the bright white lines at the grain boundaries, or 3) closely observing the magnitude of the polishing relief. A comparison of these features in the unknown with those of known ore minerals in a given polished section can provide a good value for letter hardness that may be entered into the program in addition to or in lieu of Vickers hardness number.

All letter hardness data known to the authors to be available in the literature for ore minerals have been entered into the database. The letter hardness has been estimated using a graph for those minerals whose Talmage hardness has not been reported but whose Vickers hardness number has been measured. The graph has been prepared by plotting Vickers hardness number versus Talmage hardness for 154 ore minerals.

The pre-selected search ranges for letter hardness is the plus or minus of that letter. That is, for an unknown with a letter hardness of C, the computer will search the database from C- to C+. The range of the search can be widened for the initial or for repeated searches.

If the user has not determined the letter hardness, or does not wish to use that data in the computer search of the database, he can enter "no data" for this category.

Reflectance

The database contains the mean percent reflectance in air at the standard wavelength of 546 nm for all of those ore minerals for which this value has been measured. For some ore minerals, this value has been interpolated from lists of reflectance measurements at every 20 nm. The value has been estimated for some uncommon ore minerals for which the reflectance has not yet been measured but for which the brightness has been compared to that of other ore minerals.

For those ore minerals for which the reflectance values are given in the IMA-COM data file, that information has been utilized in the ORE.ID database in preference to that available in all other sources. For those ore minerals that are not included in the IMA-COM file, interpolations of reflectance values given in Picot and Johan (1982) were used in preference to the remaining sources. Other sources used were Ramdohr (1980), Uytenbogaardt and Burke (1971), McLeod and Chamberlain (1969), Cameron (1961), Galopin and Henry (1972), and published journal articles.

For reflectance data from the IMA-COM data file (1977), Picot and Johan (1982), and other sources where a single percent reflectance value has been given for each isotropic mineral, those values were used in the database. For data on isotropic minerals from other sources, and for anisotropic mineral data from all sources, the mean reflectance has been used in the database. Thus, the mean value used in the database is the average between the minimum and maximum reflectance of a mineral where data come from some sources. Where the data come from other sources, the mean also involves the factor of interlaboratory variance. Although this involves a somewhat undesirable aspect for some ore minerals, it is a necessary procedure until precise standardized data is available for all ore minerals.

The pre-selected ranges for the computer search of the database for reflectance values vary with the reflectance of the unknown mineral. The search ranges for the various reflectance ranges are:

±5	45% - 100%
±3	35% - 44%
±5	20% - 34%
±3	15% - 19%
±5	1% - 14%

Thus, for an unknown ore mineral with a measured reflectance of 40%, the computer will search for minerals with reflectance values between 43% and 37%. Those minerals will be selected that have mean reflectance values within that range and that have other properties that also match those of the unknown.

The program has the option to change the above ranges for either the initial search or for subsequent repeated searches. If, for example, the operator has performed precise reflectance measurements in which he has strong confidence, he may wish to narrow the search range in order to reduce the number of possible minerals found by the computer program. On the other hand, if the measurements of reflectance are poor or only estimates, the operator can widen the search range to ascertain that the unknown mineral is included in the list of possible ore minerals found by the computer program.

Although ORE.ID has been designed to take advantage of the availability of quantitative data, the authors wish to emphasize that the program can be equally used by those who lack the necessary equipment for these measurements. Those individuals can obtain close estimates of reflectance by comparing the brightness of the unknown mineral with that of the known ore minerals that are present in the same polished sections. Such estimates usually can be made at least within five percent of the actual reflectance for an unknown ore mineral. Furthermore, the search ranges can be widened for searches for those minerals on which the reflectance estimates were less precise. Although the reflectance figures in the database are for a wavelength of 546 nm, those made in white light will be sufficiently close for estimation purposes.

If the user has no estimate of reflectance or does not wish to consider reflectance in the computer search of the database, he may enter "no data" for this category.

Additional reflectance data may be entered into the ORE.ID database at a future date. Such information could include reflectance at a wave length of 589 nm, reflectance values in oil, maximum and minimum reflectance for anisotropic minerals, and the quantitative color coordinates that are calculated from spectral reflectance measurements.

Strength of Anisotropism

Minerals in the database are categorized into seven groups based upon their strength of anisotropism. These are: 1) isotropic, 2) very weak, 3) weak, 4) distinct, 5) strong, 6) very strong, and 7) extreme. The pre-selected search range for strength of anisotopism is one group to either side of that determined for the unknown ore mineral. Thus, if the user assesses the anisotropism of the unknown to be strong, the computer will search the database for ore minerals with strenths of anisotropism that vary from distinct, through strong, to very strong. For those cases where the user is certain of the assessment of the strength of anisotropism, the search range may be narrowed to zero. This will result in reducing the number of possible minerals found by the computer. In contrast, should the user have little confidence in his assessment of the strength of anisotropism, he could further broaden the search range. The user can even enter "no data" in this category if he cannot assess the degree of anisotropism or if he wishes the computer to disregard anisotropism in the search of the database.

Color of the Ore Mineral

The colors of the ore minerals are contained in the database in the form of two entries: 1) the color of the ore mineral and 2) the tint shown by the ore mineral. Both the colors and the tints of the ore minerals have been grouped in the database into six categories: 1) white, 2) grey, 3) yellow, cream, orange, 4) green, 5) blue or violet, and 6) pink, red, brown, or bronze. The pre-selected search range for both the color and the tint is zero. For ore minerals with no tint, "no data" can be entered for tint. For either the color or the tint, "no data" can be entered if the user does not wish color or tint to be considered in the search of the database.

Color of the Internal Reflections

The color of the internal reflections is contained in the database in the form of two entries: 1) the color of the internal reflections, and 2) the tint of the internal reflections. The colors and the tints of the internal reflections have been grouped in the database into the same six categories as given for ore mineral color in the above section. A symbol for "no data" can be entered for those ore minerals that lack internal reflections, that have internal reflections without a color tint, or for those unknown ore minerals for which the user does not wish to consider the color and the tint of the internal reflections in the computer search of the database.

The Search

After the measured or determined data for the unknown ore mineral has been entered into the computer for each of the eight categories discussed above or for part of those categories, the computer will search the database of 573 ore minerals and find those ore minerals with properties that match the unknown. During this search the computer will display on the screen the data that have been entered and the ranges that are being searched. For example, the screen will show that the user has entered a VHN of 500 and that the computer is searching for minerals in the range from 450 to 550 VHN, that a reflectance of 40% has been entered and that the reflectance range between 37% and 43% percent is being searched, etc.

The length of time required for the search is between about 50 seconds and two minutes, depending upon the data determined for the unknown ore mineral and the search ranges selected. The main determinant of the time required to complete a search is the value entered for the color. As the database is indexed according to color, a search will execute faster if a value other than zero (for "no data") is entered. Furthermore, the search for the identity of an unknown mineral with a green color will execute faster than that for an unknown with a white color due to the fact that fewer green minerals than white minerals exist in the database.

The Data Output

The minerals with a matching property or properties are displayed on the computer screen and also may be printed. In addition to finding possible minerals to fit the unknown, the program also displays or prints out for each mineral found three other types of data: 1) the number of data matches, 2) the properties that matched, and 3) the chemical constituents.

Each of these three areas of data aid in further narrowing the number of minerals possible for the unknown ore mineral. A scan of the screen for the number of data matches for each of the ore minerals found by the computer search serves to quickly narrow the list by eliminating those that have few data matches. The list can also be narrowed by examining the specific properties that matched for each of the ore minerals found by the computer. Finally, if any information is available regarding the chemical composition of the mineral from electron microprobe analyses, SEM-EDS, or from other sources, these data can be used to further screen the list of possible ore minerals found by the computer and shown on the screen.

The final objective of the program is to select three or four ore minerals that have properties sufficiently similar to those of the unknown so that they may be considered further in descriptive texts such as Ramdohr (1980), Picot and Johan (1982), and Uytenbogaardt and Burke (1971) or published journal articles. The user will find that for many instances a single ore mineral that best matches the data for the unknown will be evident from the computer search together with the number and character of data matches and chemical constituents that are displayed on the screen.

Repeated Searches

Additional computer searches of the database with the previously entered data for an unknown ore mineral may be undertaken without reentry of that data. Changes in data input or search ranges may be limited for those selected categories while the remaining data entered for the initial search can be retained for repeated searches.

APPLICATIONS OF ORE.ID TO ORES AND BENEFICIATION PRODUCTS

The following are some examples that have been selected to illustrate the application of ORE.ID to the identification of ore minerals that occur in ore deposits and beneficiation products. Each of the examples represent applications to actual case histories.

Mahd adh Dhahab, Saudi Arabia gold ores

The Mahd adh Dhahab gold mine was the largest gold producer among numerous former gold mines in the Precambrian Shield near the Red Sea in southwestern Saudi Arabia. During an ore microscopic examination of gold ores from the Mahd adh Dhahab gold mine in Saudi Arabia, grains of a white mineral with a violet tint were observed to occur in veinlets of native silver, native gold, and tetrahedrite that traverse sphalerite, galena, and chalcopyrite grains. The grains of the violet white mineral were isotropic, had a Vickers hardness number of about 55, a Talmage hardness of A+, and reflectance of 37%. The input of these data into the ORE.ID program resulted in the following output:

6 Petzite (Ag,Au,Te) VHN, LH, Ref, Iso, Col, Tint

The results from this case history indicated that the mineral was petzite, that six of the categories in the database for petzite matched those of the unknown mineral, and that the data for petzite that matched petzite were Vickers hardness number (VHN), letter hardness (LH), reflectance (Ref), degree of anisotropism (Iso), color (Col), and tint (Tint). The fact that petzite contains silver and gold and that native silver and native gold are phases associated with the unknown are further support for the identification of this ore mineral.

Southeast Missouri Lead Ores

The Southeast Missouri Lead District is the world's largest lead producing district. Although the principal sulfide minerals in the district are galena, sphalerite, marcasite, pyrite, and chalcopyrite, small amounts of other ore minerals are locally present. A recent ore microscopic examination of core obtained during an exploration program revealed the presence of minor grains of a mineral uncommon to the ores of that district. Measurements of Vickers hardness number for the mineral yielded 365, its Talmage hardness was determined to be D-, and the average measured reflectance was 26%. The mineral was orange with a pink tint, had very strong anisotropism, and it lacked internal reflections. With the input of these data into the ORE.ID program, the computer search found a single mineral, luzonite, with the following output data:

1 Billingsleyite Ref
6 Luzonite (Cu,As,S) VHN, LH, Ref, Iso, Col, Tint
1 Routhierite (Tl,Cu,Ag,Hg,Zn,As,Sb,S) Ref

Although three ore minerals were found by the computer search, billingsleyite and routhierite can be eliminated from further consideration because only one data item for each of those minerals match that of the unknown. In contrast, six of the properties in the database for luzonite matched that of the unknown mineral.

Creta Copper Mine, Southwest Oklahoma

The Creta copper mine in southwest Oklahoma was one of the most productive of the red bed sandstone- and shale-hosted copper deposits in the southwestern part of the United States. At one time it reached a production rate of about 1000 tons of ore per day and had an average mill head grade of about 2% copper. Ore microscopic study of the Creta ores and concentrates has shown that copper sulfide grains occur predominantly as replacements of megaspores, colloform pyrite, and pyrrhotite crystals (Hagni and Gann, 1976). During continuing ore microscopic studies of the Creta ores (Hagni and Gann, this volume), grains of an isotropic mineral that was white with a very faint pink tint were found to be locally present in these copper ores. Quantitative measurements on the unknown gave a reflectance of 43% and a Vickers hardness number of 80. The mineral had an estimated Talmage hardness of B and lacked internal reflections. A computer search of the ORE.ID database produced the following results:

3	Bohdanowiczyte (Ag,Bi,Se)	VHN, LH, Ref
6	Galena (Pb,S)	VHN, LH, Ref, Iso, Col, Tint
5	Guadalcazarite (Hg,Zn,S,Se)	VHN, LH, Iso, Col, Tint

Among the three ore minerals found by the computer search, bohdanowiczyte can be disregarded because of its small number of data matches. Thus, in this case history, two possible ore minerals were found, galena and guadalcazarite, and these minerals should be considered further by reference to standard volumes on the properties of ore minerals. Additional study of such volumes shows that guadalcazarite is not a likely possibility for the unknown because it has a brown color tint (brown and pink color tints are grouped together in the ORE.ID database) and because that mineral is typically found in association with other mercury minerals, such as cinnabar. Finally, those two possible minerals can be distinguished by some type of chemical analysis. Subsequent analyses by scanning electron microscope - energy dispersive spectroscopic analysis (SEM-EDS) confirmed that the mineral contained only lead and sulfur and thus was galena.

Although the Creta copper ores contained recoverable amounts of silver, the silver-bearing mineral was only discovered recently (Hagni and Gann, this volume). A gray mineral that yielded measurements of about 30 for Vickers hardness number and a mean reflectance of 28% was found to be locally intergrown with chalcocite. It can be distinguished with difficulty from the chalcocite-group minerals by its faint violet tint, but it is best recognized by its strong anisotropism. A letter hardness of B was determined for the mineral by observing scratches that crossed the border between its grains and those of djurleite and chalcocite. The mineral lacked internal reflections. These data were input to the microcomputer program with the following results:

1	Billingsleyite	Ref
5	Orpiment (As,S)	VHN, LH, Ref, Iso, Col
4	Ottemannite (Sn,S)	VNN, Ref, Iso, Col
1	Routhierite (Tl,Cu,Ag,Hg,Zn,As,Sb,S)	Ref
6	Sabatierite (Cu,Tl,Se)	VHN, LH, Ref, Iso, Col, Tint
6	Stromeyerite (Cu,As,S)	VHN, LH, Ref, Iso, Col, Tint
6	Weissite (Cu,Te)	VHN, LH, Ref, Iso, Col, Tint
1	Zirconolite (Ca,Zr,Ti,O)	Col

The five minerals with a sufficient number of data matches that were found by the computer, orpiment, ottemannite, routhierite, sabatierite, stromeyerite, and weissite should be considered for additional literature study. Such study shows that orpiment can be eliminated from further consideration due to its strong white to yellow internal reflections, ottemannite is eliminated by its orange brown internal reflections, sabatierite has a much stronger blue color than the unknown, and weissite is eliminated by its weak to moderate anisotropism (alternatively, a rerun of the program with the search range for anisotropism restricted to strong eliminates weissite from the list of possible minerals). Subsequent analyses by SEM-EDS and by electron microprobe confirmed that the mineral was stromeyerite (Hagni and Gann, this volume).

Silver Mine District, Missouri

A recent ore microscopic study of the silver-tungsten ores that occur in the Silver Mine District in southeastern Missouri revealed the presence of nine minerals that were not previously known to occur in that district (Hagni, in press). Although the silver content of those ores was previously believed to occur in solid solution in galena, three separate silver-bearing minerals were detected by ore microscopy. For one of those minerals the determined properties were: VHN = 325, Talmage hardness = D, reflectance = 30%, isotropic, grey color with a blue tint, and rare red internal reflections with brownish tints. A computer search of the ORE.ID database for minerals with these properties produced:

2	Avicennite (Tl,O)	Iso, Col
1	Billingsleyite	Ref
3	Chameanite	Ref, Iso, Col
2	Dzhalindite (In,O,H)	Iso, Col
5	Freibergite (Cu, Fe, Zn, Ag, Sb, S)	VHN, LH, Ref, Iso, Col
2	Monteponite (Cd,O)	Iso, Col
1	Routhierite (Tl,Cu,Ag,Hg,Zn,As,Sb,S)	Ref
6	Tennantite (Cu,Fe,As,S)	VHN, LH, Ref, Iso, Col, Tint

The two minerals with a sufficient number of data matches are the two fahlores, freibergite and tennantite. Although the individual fahlores often can be distinguished by their color tints, their determination is best accomplished by accompanying SEM or electron microprobe analyses. Subsequent electron probe analyses of this mineral showed that it was argentiferous tennantite (Hagni, in press).

A second silver mineral, one that was found to be closely associated with argentiferous tennantite in the Silver Mine ores, exhibited a strong reflective pleochroism from greenish grey to violet grey, strong anisotropism, and no internal reflections. Talmage hardness was determined to be C. Quantitative measurements of Vickers hardness number and reflectance yielded 95 and 29.4%. A computer search for minerals with these properties yielded three minerals, none of which had a significant number of data matches. The three minerals found were:

1	Billingsleyite	Ref
1	Routhierite (Tl,Cu,Ag,Hg,Zn,As,Sb,S	Ref
1	Zirconolite (Ca,Zr,Ti,O)	Col

Anticipating that Vickers hardness might be the problem property, the program was rerun with the same input data except for an entry of "no data" for VHN. The computer search found thirteen possible ore minerals that had some data matches for the unknown mineral.

1	Billingsleyite	Ref
3	Cryptomelane (K,Mn,O,H)	Ref, Iso, Col
2	Doloresite (V,O,H)	Iso, Col
4	Enargite (Cu,As,Sb,S)	LH, Ref, Iso, Col
2	Manjiroite (Na,Ca,K,Ba,Mn,Fe,Al,Mg,O)	LH, Iso
3	Ottemannite (Sn,S)	Ref, Iso, Col
4	Pearceite (Ag, Cu, As, Sb,S)	LH, Ref, Iso, Col
2	Permingeatite (Cu,Sb,As,Se)	LH, Ref
4	Polybasite (Ag,Sb,S)	LH, Ref, Iso, Col
4	Pyrargyrite (Ag,Sb,S)	LH, Ref, Iso, Col
1	Routhierite (Tl,Cu,Ag,Hg,Zn,As,Sb,S)	Ref
2	Shadlunite (Cu,Fe,Pb,Cd,S)	LH, Ref
1	Zinc (Zn)	LH

For the results of this repeated search where Vickers hardness data were not utilized four minerals that have four data matches or the six minerals with three or more data matches can be considered for additional comparison with other properties in standard reference volumes. In addition, chemical data can be sought for the unknown mineral to aid in selecting among those possible minerals.

An alternative approach to identifying the above mineral by ORE.ID is to adjust the value of the problem property. For example, if the Vickers hardness determination was obtained on fine-grained material or if the mineral exhibited some cracking around the indentation, the measured VHN may be on the soft side. A rerun of the program with the same input data, except with the VHN adjusted to 185 gives the following results:

1	Billingsleyite	Ref
3	Manjiroite (Na,Ca,K,Ba,Mn,Fe,Al,Mg,O)	VHN, LH, Iso
5	Pearceite (Ag,Cu,As,Sb,S)	VHN, LH, Ref, Iso, Col
3	Permingeatite (Cu,Sb,As,Se)	VHN, LH, Ref
1	Routhierite (Tl,Cu,Ag,Hg,Zn,As,Sb,S)	Ref
3	Shadlunite (Cu,Fe,PB,Cd,S)	VHN, LH, Ref
1	Zirconolite (Ca,Zr,Ti,O)	Col

This repeated computer search found that six categories in the database for pearceite matched those of the unknown. The data matches were for VHN, LH, Ref, Iso, and Col. Electron microprobe analyses have shown that the mineral is antimonpearceite (Hagni, in press).

A third silver mineral was found to occur independently from the grains of argentiferous tennantite and antimonpearceite in the ores from the Silver Mine district. That mineral was discovered to occur as fine-grained aggregates that replace galena and chalcopyrite. The mineral is white and appears to be very similar in its properties to galena, but it can be distinguished from that mineral by its weak to distinct anisotropism. Its Talmage hardness was estimated to be B+. Measurements of Vickers hardness number and reflectance yielded 134 and 41.3 to 41.5%. The mineral lacked internal reflections. A computer search for possible ore minerals with those properties yielded the following:

6	Berryite (Pb,Cu,Ag,Bi)	VHN, LH, Ref, Iso, Col, Tint
1	Kennedyite (Fe,Mg,Ti,O)	Iso
3	Plumbian Ikunolite (Bi,Pb,S,Se,Te)	Iso, Col, Tint

Of the three possible ore minerals found by the computer search, probably only berryite has a sufficient number of data matches to merit further study in the literature. Electron probe analyses have shown that this is true and that the unknown is the uncommon mineral, berryite (Hagni, in press).

SUMMARY AND CONCLUSIONS

The increasingly widespread use of microcomputer technology and the use of relational database management systems on microcomputers has provided the foundation for the development of an ore mineral identification system that is rapid, easy to use, flexible, utilizes most of the properties normally determined on ore minerals, and searches a database consisting of a very large number of ore minerals. The microcomputer program, ORE.ID, is the first to accomplish all of those goals and to rapidly find the ore mineral or minerals that have properties that most closely match those of a given unknown ore mineral.

ORE.ID is available in a diskette from Delta Research Corporation, 1401 Wilson Boulevard, Arlington, Virginia 22209.

ACKNOWLEDGEMENTS

The writers wish to acknowledge Delta Research Corporation for the interest and support during the preparation of the ORE.ID program and for providing one of us (J.E.H.) with the time to prepare the program and enter the database. One of us (R.D.H.) wishes to acknowledge the opportunity to study the outstanding systematic collection of ore minerals in polished sections in the Ramdohr collection at Heidelberg University during his sabbatical study as a Deutscher Akademischer Austauschdienst scholar in the spring of 1983. Some of the information on the properties of rare ore minerals that is included in the identification tables in his short course workbook and that is included in ORE.ID was obtained by determinations made on polished sections in the Ramdohr-Heidelberg collection.

REFERENCES

Atkin, B. P., and Harvey, P. K., 1979a, "Nottingham Interactive System for Opaque Mineral Identification: NISOMI," Institute of Mining and Metallurgical Transactions, London, Vol. 88, pp. B 24-27.

Atkin, B. P., and Harvey, P. K., 1979b, "The Use of Quantitative Color Values for Opaque Mineral Identification," Canadian Mineralogist, Vol. 17, pp. 639-647.

Atkin, B. P., and Harvey, P. K., 1982, "NISOMI-81: An Automated System for Opaque Mineral Identification in Polished Section," pp. 77-91 in Process Mineralogy II: Applications in Metallurgy, Ceramics, and Geology, Hagni, R. D., ed., The Metallurgical Society, AIME, New York, N.Y.

Bernhardt, H.-J., 1979, "Computer-Supported Ore Mineral Identification using Spectral Reflectance in Visible Light," Neues Jahrbuch fur Mineralogie, Monatshefte 9, pp. 403-407.

Bowie, S. H. U., and Simpson, P. R., 1978, The Bowie-Simpson System for the Microscopic Determination of Ore Minerals: First Students Issue, London, 10 pp., 2 tables and 5 charts.

Bowie, S. H. U., and Taylor, K., 1958, A System of Ore Mineral Identification, Mining Magazine, London, Vol. 99, pp. 265-277 and 337-345.

Cameron, E. N., 1961, Ore Microscopy, Wiley, 293 pp.

Chamot, E. M., and Mason, C. W., 1940, Handbook of Chemical Microscopy, Vol. II, Wiley, New York, 438 pp.

Craig, J. R., and Vaughan, 1981, Ore Microscopy and Ore Petrography, Wiley, New York, 406 pp.

Davy, W. M., and Farnham, C. M., 1920, Microscopic Examination of the Ore Minerals, McGraw-Hill Book Co., New York, 154 pp.

Eales, H. V., 1969, "The Bases of a Scheme for Identification of Ore Minerals by Optical Methods," Mineralium Deposita, Vol. 4, pp. 52-64.

Farnham, C. M., 1931, Determination of the Opaque Minerals, McGraw-Hill Book Co., New York, 236 pp.

Fairbanks, E. E., 1946, "The Punched Card Identification of Ore Minerals," Economic Geology, Vol. 41, pp. 761-768.

Galopin, R., and Henry, N. F. M., 1972, Microscopic Study of Opaque Minerals, Heffner, 322 pp.

Hagni, R. D., 1984, Applied Ore Microscopy, Short Course Wookbook, University of Missouri-Rolla, 731 pp.

Hagni, R. D., in press, "Ore Microscopy of the Silver Minerals in the Epigenetic Ag-W-Sn Deposits in the Silvermine District, Southeastern Missouri, U.S.A.", Syngenesis and Epigenesis in the Formation of Mineral Deposits, A. Wauschkuhn, ed, Springer-Verlag, Heidelberg.

Hagni, R. D., "Precambrian Silver-Tungsten-Tin Mineralization and its Relationship to Granitic Magmatism in Southeastern Missouri, U.S.A.", VI International Association on the Genesis of Ore Deposits Symposium, Tbilisi, U.S.S.R., in press.

Hagni, R. D., and Gann, D. E., 1976, "Microscopy of Copper Ore at the Creta Mine, Southwestern Oklahoma", Oklahoma Geological Survey Circular 77, pp. 40-50.

Hagni, R. D., and Gann, D. E., "Ore Microscopy and Genesis of the Copper-Shale Deposit at Creta, Oklahoma", this volume.

Henry, N. F. M., 1977, IMA-COM Quantitative Data File: First Issue, McCrone Research Associates, London, 170 cards.

Kühnel, R. A., Prins, J. J., and Roorda, H. J., 1980, The Delft System for Mineral Identification: Opaque Minerals, Delft University Press, Delft, 204 pp.

Leonard, B. F., 1979, "Index to the Reflectance and Microindentation Hardness of Ore Minerals in the IMA/COM Quantitative Data File (First Issue, 1977)," United States Geological Survey Open-File Report 79-658, 52 pp.

McLeod, C. R., and Chamberlain, J. A., 1969, Reflectivity and Vickers Microhardness of Ore Minerals - Charts and Tables, Geological Survey of Canada, Paper 68-64.

Murdoch, J., 1916, Microscopic Determination of the Opaque Minerals: An Aid to the Study of Ores, Wiley, New York, 165 pp.

Picot, P., and Johan, Z., 1977, Atlas of Ore Minerals (in French), Science and Medicine, Pubications of France Corporation, New York, N.Y., 408 pp.

Picot, P., and Johan, Z., 1982, Atlas of Ore Minerals (in English), Bureau de recherches geologiques et mineres, Cedex, and Elsevier, Amsterdam, 458 pp.

Ramdohr, P., 1980, The Ore Minerals and their Intergrowths, 2nd English edition, Pergamon Press, Oxford, 1205 pp.

Schneiderhöhn, H., 1952, Erzmikroskopisches Praktikum, E. Schweizerbart'sche Verlagsbuchhandlung, Stuttgart, 274 pp.

Short, M. N., 1940 (reprinted 1948), Microscopic Determination of the Ore Minerals, United States Geological Survey Bulletin 914, 314 pp.

Tarkian, M., 1974, A Key-Diagram for the Optical Determination of Common Ore Minerals, Minerals Sci. Engineering, Vol. 6, pp. 101-105.

Uytenbogaardt, W., and Burke, E. A. J., (1971), Tables for Microscopic Identification of Ore Minerals, Elsevier, 430 pp.

Van der Veen, R. W., 1925, Mineragraphy and Ore-Deposition, Vol. I, G. Naeff, The Hague, 168 pp.

Chapter 5

THE APPLICATION OF CATHODOLUMINESCENCE MICROSCOPY TO THE STUDY OF SPARRY DOLOMITE FROM THE VIBURNUM TREND, SOUTHEAST MISSOURI

by Robert L. Voss and Richard D. Hagni

Mine Geologist, Cominco American Inc.
Bixby, MO

Gulf Oil Foundation Professor of Geology, University of Missouri-Rolla
Rolla, MO

ABSTRACT

Sparry dolomite from all of the operating mines in the Viburnum Trend of Southeast Missouri have been found by cathodoluminescence microscopy to reveal a systematic pattern of banding resulting from a growth zoning within individual dolomite rhombs. Four principal bands have been recognized throughout the entire Trend and each of the major bands contain even finer internal banding. Minor bands a few microns in width can be correlated in specimens collected ten's of kilometers apart along the trend. The first three principal dolomite bands decrease from south to north. This feature, together with the fact that the white sparry dolomite crystals decrease in abundance from south to north in the Trend, suggests that a major source for the fluids was from the south. The application of cathodoluminescence microscopy to crystalline dolomite demonstrates a strong correlation between ore grade and the intensity of internal banding and dissolution textures.

INTRODUCTION

The gangue dolomite crystals that occur with the sulfide minerals in the ore deposits of the Viburnum Trend in the Southeast Missouri Lead District have been largely neglected in previous mineralogical studies. Intensive examinations of the paragenetic sequence shown by the sulfide minerals have indicated that most aspects of the sequence are present in the ores at all of the mines (Hagni and Trancynger, 1977; Jessey, 1981, Rickman, 1981; Horrall et al., 1983). These observations have suggested that at any given time the same or similar ore fluids may have permeated a large part or the entire district (Hagni, 1983).

Studies in other Mississippi Valley-type districts have shown that thin depositional zones within single crystals can be correlated for long distances. Three major zones were detected within growth zoned sphalerite crystals in the Wisconsin-Illinois lead-zinc district, and many of these were correlated over a distance of 23 km (McLimans, Barnes, and Ohmoto, 1980). As many as six growth zones were found by cathodoluminescence microscopy to occur within single dolomite crystals in the East Tennessee zinc district, and these were traced throughout that district by Ebers and Kopp, 1979. The fact that such correlations can be made throughout those districts strongly suggests that the ore fluids that deposited those minerals were pervasive throughout each of those two districts.

In the course of a study of mineral paragenesis and chemistry of deposition of the ores at the Milliken mine, Rickman (1981) found that four growth zones could be distinguished within single crystals of sparry dolomite by cathodoluminescence microscopy. Three of these zones were traced by Braunsdorf and Lowman (1983) eastward from the Milliken mine into the backreef facies. Carbon and oxygen isotopic compositions were measured for each of the zones.

The purpose of the present investigation has been to determine the lateral extent to which the zones first recognized by Rickman might be traced throughout the Viburnum Trend. Sparry dolomite was selected for this study because it is the most abundant gangue mineral in the Viburnum ores and because its close relationship with the sulfide minerals has been demonstrated by many previous investigations, including the recent studies by Rickman (1981), Mugel (1983), Horrall (1983), and Anderson (1983).

The study was initiated with an intensive sampling of dolomite thoughout the Milliken mine in order to first establish whether the four zones could be recognized throughout a single mine and to determine if their character was dependent upon stratigraphic or structural position. Subsequently the study was enlarged to examine dolomite crystals from all of the operating mines in the Viburnum Trend, and a total of 100 polished thin sections were carefully examined by cathodoluminescence microscopy. This paper describes the distinguishing characteristics of the four growth zones in sparry dolomite, provides electron microprobe analyses for each of the zones and some sub-zones, relates the corrosion characteristics of the surfaces of individual zones to dolomite dissolution associated with the ore deposits, and interprets the results in terms of extent and direction of ore fluid movement in the Viburnum Trend.

Location

The Viburnum Trend is located in southeastern Missouri approximately 30 kilometers west of the St. Francois Mountains. The Trend is the largest subdistrict within the larger Southeast Missouri Lead

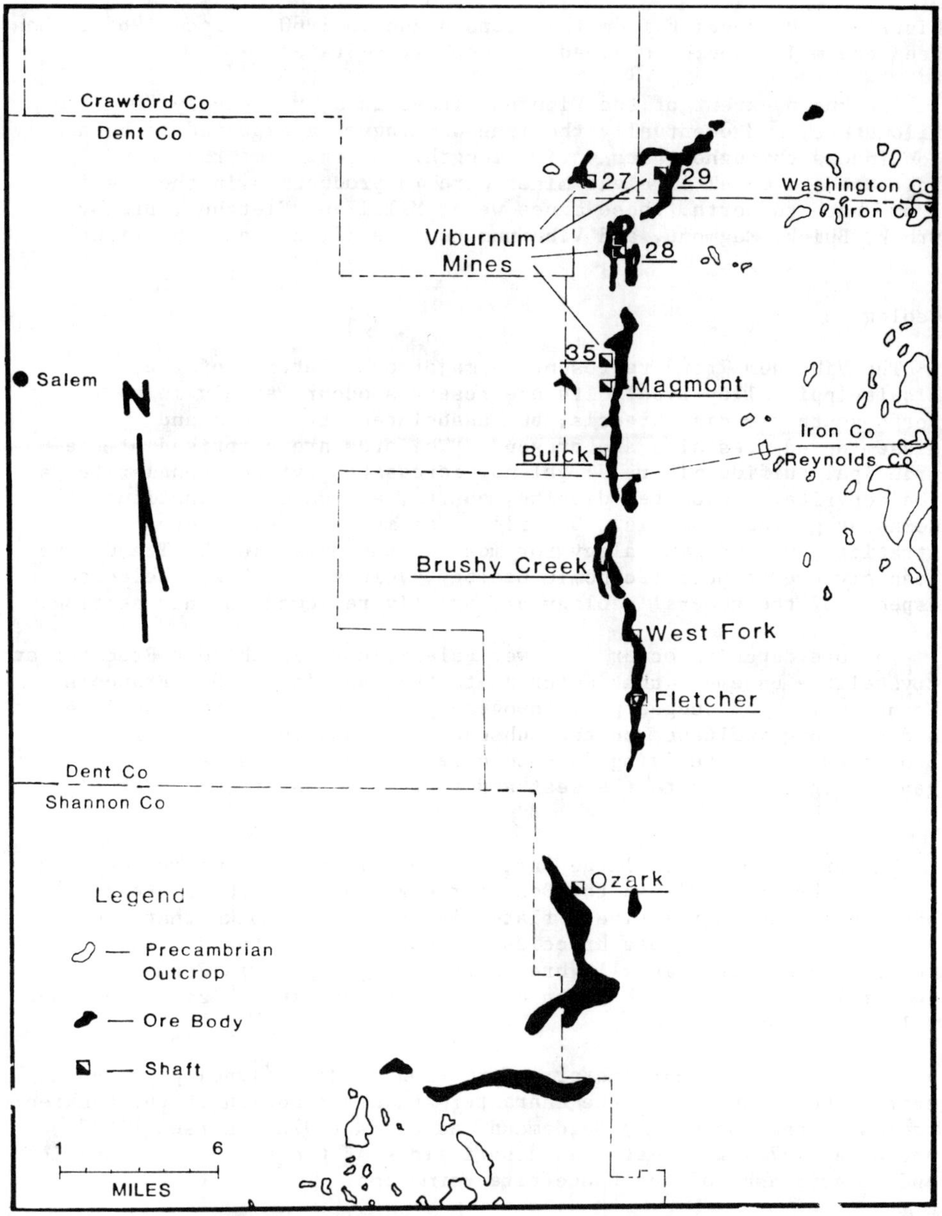

Figure 1. Map showing the distribution of ore deposits, positions of Precambrian outcrops, and locations of mine shafts in the Viburnum Trend, Southeast Missouri Lead District. Mines from which samples were obtained for this study are underlined.

District. Production from the Trend began in 1960. Since 1965 it has been the major source of lead in the United States.

The known extent of the Viburnum Trend is approximately 70 kilometers, and eventually the mine workings are expected to be nearly continuous throughout its entire length. At the time this study was initiated, a total of eight mines were in production in the Trend. From south to north, those mines were: Milliken, Fletcher, Brushy Creek, Buick, Magmont, and Viburnum numbers 35, 28, and 29 (Figure 1).

Geology

The Viburnum Trend represents a major ore district of the Mississippi Valley-type. Its ore reserves occur largely in a north-south trending breccia, but associated stratiform and disseminated ores also are present. The ores are comprised of the principal sulfide minerals: galena, marcasite, pyrite, sphalerite, and chalcopyrite. Dolomite, dickite, quartz, and calcite constitute the dominant gangue minerals. Detailed treatments of structure, stratigraphy, and mineralogy for most of the mines in the Trend have been given elsewhere (Economic Geology, Volume 72, 1972); selected aspects of the general geology are briefly reviewed in this section.

The ore deposits occur in lower Paleozoic rocks above a Precambrian crystalline basement that outcrops to the east in the St. Francois Mountains. The rugged paleotopography of the Precambrian surface has had a strong influence on the subsequent Paleozoic sedimentary processes. The overlying Paleozoic sedimentary rocks have a very gentle regional dip to the west and southwest away from the St. Francois Mountains.

The ore deposits in large part are localized in slump, slide, and collapse breccias within the Bonneterre Formation. These breccias have developed in relatively flat-lying carbonate rocks that are now dolomites. The collapse breccias are bounded by outward dipping reverse faults that locally breach the overlying Davis Formation. Temporal relationships between brecciation and mineralization are not fully resolved.

Early work on the Bonneterre Formation by Howe (1968) provided a better understanding of the character and distribution of the backreef facies. Recent papers by Gerdemann and Myers (1972), Larsen (1977), and Lyle (1977) give detailed descriptions of the facies, diagenesis, and stratigraphy of the Bonneterre Formation.

The Bonneterre Formation has been divided into Lower, Middle, and Upper units. The base of the Bonneterre Formation lies conformably on the Lamotte Sandstone, and is marked by a sandy transition zone. The remaining lower Bonneterre is composed of interbedded digitate stromatolites and calcarenites. A thick section of digitate stromatolite reef usually caps the lower Bonneterre. The lower Bonneterre represents the backreef and reef facies. The middle Bonneterre is composed of medium crystalline pelletal and oolitic dolomite that represent the fore reef facies. The upper Bonneterre shows basinal features and is composed of fine siltstones, brown calcarentite, and dark green shale bands. The 280 foot average thickness of the Bonneterre shows an overall transgressive sequence deposited on a restricted carbonate platform (Lyle, 1977; Larson, 1977). Ore grade mineralization is mined from all three portions of the Bonneterre Formation.

Method of Study

Samples for this investigation were collected from all producing mines in the Viburnum Trend. In so far as possible, the sample sites were selected to provide a representative areal distribution within each mine. The Milliken or Ozark Lead mine, at the south end of the Trend, was sampled intensively during the initial stages of the study to establish whether a consistent pattern of dolomite zones could be traced throughout a single mine and to determine whether any variations in the dolomite zones were related to stratigraphic ore horizons or different structural settings. The stope numbers, stratigraphic positions, and structural settings were recorded for each collected specimen.

Highly polished thin sections were prepared from each specimen and these were carefully examined by cathodoluminescence microscopy using a Nuclide Corporation Luminoscope. The most common Luminoscope settings employed were 0.8 milliamps, 12 killovolts, and a spot diameter of 1 cm. Cathodoluminescence photomicrographs were taken from each polished thin section using a Zeiss M63 automatic photometer and a Zeiss 35 mm camera. Various types of film were tested. The best results were obtained using Ektachrome 400 ASA and 3M 640 ASA films that were exposed and developed at 1600 ASA and 1280 ASA respectively. This combination of equipment and films resulted in relatively short exposure times of only about 0.25 to five minutes. Such short exposure times are very desirable because they prevent burning of the specimens and minimize operator fatigue.

CATHODOLUMINESCENT DOLOMITE ZONES

Four growth zones can be recognized with cathodoluminescence microscopy to occur in single crystals of secondary sparry dolomite. Each of the zones are readily recognized on the basis of their texture, degree of crystal development, color, presence or lack of banding, and intensity of cathodoluminescence. All four zones in the secondary dolomite crystals are readily distinguished from the host-rock dolomite by their cathodoluminescence characteristics (Figure 2).

The host dolomite is moderately cathodoluminescent to non-cathodoluminescent. Under ordinary plane polarized light the hostrock dolomite contains numerous small brown inclusions of organic material and exhibits relic textures that correspond to the original rock type. Oolites, intraclasts, pelletoids, and fossil fragments are common. Under cathodoluminescence the host dolomite usually is non-luminescent, but it may show many irregularly shaped brightly luminescent spots. Individual grains in the host dolomite are difficult to distinguish under cathodoluminescence except where secondary sparry dolomite overgrowths delineate grain boundaries.

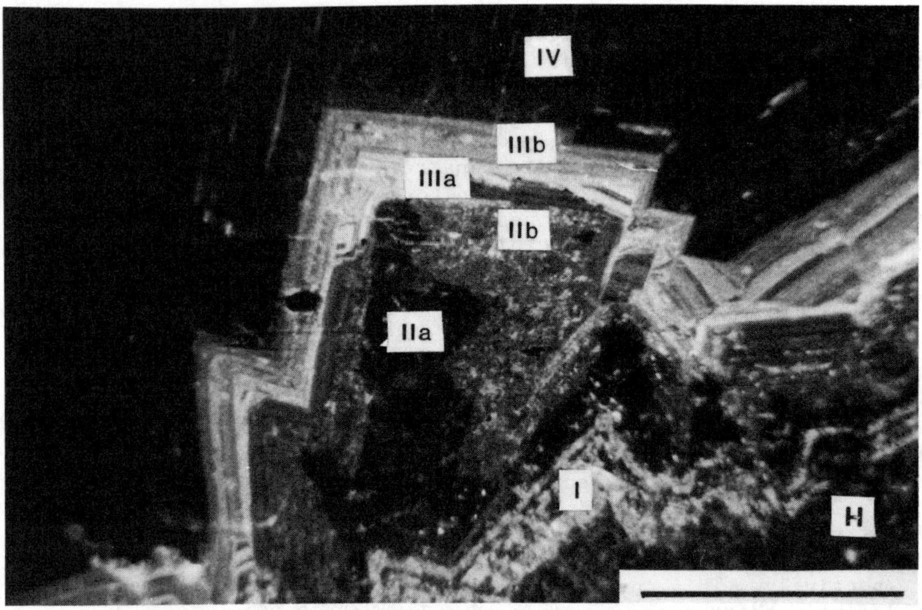

Figure 2. Cathodoluminescence photomicrograph showing a typical dolomite rhomb from the Milliken Mine, Viburnum Trend, Southeast Missouri Lead District. H = host dolomite The four major zones and subzones are shown. Scale bar = 0.5 mm

Grains of detrital potassium feldspar are readily recognized by their bright blue cathodoluminescence. They range from 100 to less than 10 micrometers and average less than 20 micrometers, and they are irregular in shape. Some of the larger grains, which are nearly 100 micrometers across, exhibit rounding. The feldspar grains are commonly found in the host dolomite. Shale seams and stylolites contain especially dense quantities of blue cathodoluminescent feldspar. In contrast, blue luminescent feldspar grains are not found within the four generations of late sparry dolomite.

Zone I

The earliest zone deposited within sparry dolomite crystals in the Viburnum Trend is designated "zone I" in this paper, and can be recognized under plane polarized light by its cloudy core. The cloudy core is the result of the abundant fluid inclusions that are contained within the dolomite in this zone (Rowan, 1984). Zone I dolomite can be recognized in hand specimens as a milky white dolomite druse that lines vugs. Examination of these specimens by cathodoluminescence microscopy showed that the vug-filling dolomite consisted almost entirely of zone I spar (Figure 4e). Zone I cement may be an important factor in limiting the porosity and thereby the flow of later ore depositing fluids. In addition, all stages of conversion of the substrate to zone I dolomite are present, and specimens in which the hostrock has become completely converted to zone I have been observed. As an example, in one specimen of the bleached algal reef from the Viburnum 29 mine, zone I characteristics were superimposed upon the entire specimen.

Zone I can be summarized as having the following characteristics: a moderate to bright cathodoluminescence, mottled texture, a distinct absence of well-defined banding, and a gradational contact with the substrate dolomite. Zone I exhibits greater variablity in its cathodoluminescence characteristics than the other three zones. This variability may be the product of the original lithologies of the hostrock. Zone I is pre-ore and examples of sulfides replacing the earliest generation of dolomite are discussed in a subsequent section.

Zone II

Zone II can best be described as non-cathodoluminescent. It usually is deposited on the corroded or rounded surface of the underlying zone I. In most dolomite crystals, zone II occurs as a thin dark band that marks the boundary between zones I and III. Zone II occasionally is slightly cloudy or shows faint internal banding.

In three of the mines studied, Fletcher, Brushy Creek, and Buick, a very thin, bright cathodoluminescent marker band occurs within zone II. This zone is easily recognized and was traced for a distance of sixteen kilometers along the length of the Viburnum Trend.

Zone III

Zone III is the most readily recognizable because it is characterized by numerous bands of alternating bright and non-cathodoluminescent spar. Zone III also has the brightest catholuminescence of the four dolomite generations. Dissolution textures are always present and readily recognized within zone III.

More than 60 individual bands were counted within zone III. The areal correlation of the individual bands is very difficult, but small groups of bands can be recognized from one mine to another. In those specimens where the dissolution textures are not well developed, zone III can be subdivided into two sub-zones, III-a and III-b. Zone III-a is brightly cathodoluminescent and zone III-b displays moderate to bright cathodoluminscence.

Some small dolomite rhombs are composed entirely of zone III. These usually can be recognized in hand specimens as grey sparry dolomite crystals.

Zone IV

Zone IV dolomite is dull to non-cathodoluminescent and always forms euhedral crystal faces. Faint banding can be detected in some dolomite crystals. A distinct correlation between zone IV and saddle dolomite development was recognized. For example, dolomite crystals that are large, pink or white, and saddle-shaped are revealed by cathodoluminescence microscopy to consist almost entirely of zone IV dolomite. Zone IV dolomite was deposited after most of the ore deposition, except for some late chalcopyrite and pyrite crystals.

DISSOLUTION TEXTURES

Dissolution or corrosion textures were observed to varying degrees in all of the specimens examined for this study. Two major types of dissolution textures were recognized and they are termed "passive" and "dynamic". Passive dissolution is characterized by subhedral to anhedral crystal faces. This type of dissolution was a relatively early event that occured after the deposition of zone I dolomite and prior to the crystallization of zone II dolomite. No brecciation event has been directly correlated with passive dissolution. That is, zone I is never brecciated by itself, but rather brecciation that involves zone I is of the dynamic type dissolution that is related to zone III deposition.

Dynamic dissolution occurs after zone II and prior to the deposition of zone IV. Dynamic dissolution shows various textures, such as convoluted, truncated, and fractured bands, rubble zones, and angular unconformities with the underlying and overlying bands. These textures that were formed by dynamic dissolution should not be confused with those of pseudofaces described by Walkden and Berry (1984).

Breccias have been recognized throughout the Trend. The breccias have provided a permeable conduit for ore fluids and served as the focus for deposition of sulfide and gangue minerals. Dolomite spar collected from the breccias shows intense dynamic dissolution, displaying such features as convoluted and truncated banding, fractured and chip crystals, and rubble zones within individual crystals. Specimens collected from the breccia ore bodies show more internal bands within zone III than those collected form the lower grade ore bodies.

Several features that commonly are observed in hand samples and in thin sections show a close temporal relationship between brecciation (dolomite dissolution) and zone III deposition. One of these involves fractures that commonly were observed to be filled with dolomite belonging only to zone III. These dolomite-filled fractures crosscut the earlier zones I, II, and the host dolomites (Figure 3a). Another feature is that of breccia fragments that exhibit four generations of dolomite growth on three sides, but have only zone III deposition on the fourth side. Many examples of these relationships were found during this cathodoluminescence investigation. These features indicate that a major brecciation event occurred during or just prior to zone III deposition.

VARIATIONS IN DOLOMITE ABUNDANCE AND MICROSTRATIGRAPHY ALONG THE TREND

The examination of about one hundred polished thin sections by cathodoluminescence microscopy has revealed a correlatable and temporal banding throughout the Viburnum Trend. To illustrate this correlation, representative specimens from six mines from the south to north end of the Trend are shown in Figure 4. All four of the major dolomite zones are clearly visible in each of the specimens. Several trends have been determined from observations of the banding and dolomite abundance.

The abundance of white sparry dolomite gangue crystals varies significantly within the Viburnum Trend showing a marked decrease from south to north. The ores at the Milliken and Fletcher mines, where sparry dolomite gangue occurs in the greatest abundance, have large vugs and breccia clasts that commonly are coated with this mineral. In contrast, to the north at the Brushy Creek, Buick, and Magmont mines, the occurrence of sparry dolomite is largely limited to individual crystals or as small clusters of crystals in vugs. With rare exceptions, sparry dolomite was not observed as a drusy coating in vugs or on breccia clasts in those three mines. White sparry dolomite is rare at the Viburnum 29 mine at the north end of the Trend where only a few individual crystals were found during the sampling of

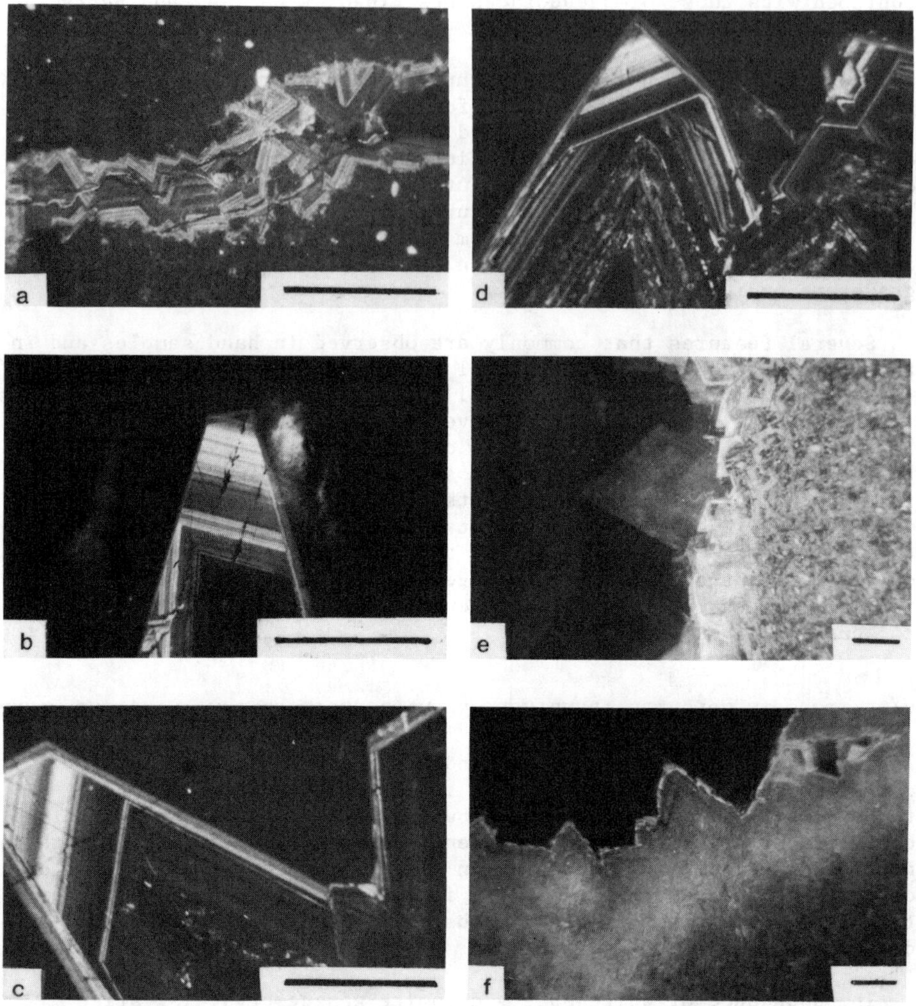

Figure 3. Cathodoluminescence photomicrographs of Viburnum Trend dolomite. a - fracture filled with zone III only, bright spots are potassium feldspar, b, c, and d show close correlation among internal bands in zone III. b - Milliken mine, c - Fletcher mine, and d - Buick mine; e shows large zone IV overgrowing small zone III crystal from the Buick mine; f shows the large zone I component of milky white spar - Magmont mine. Scale bar = 250 micrometers.

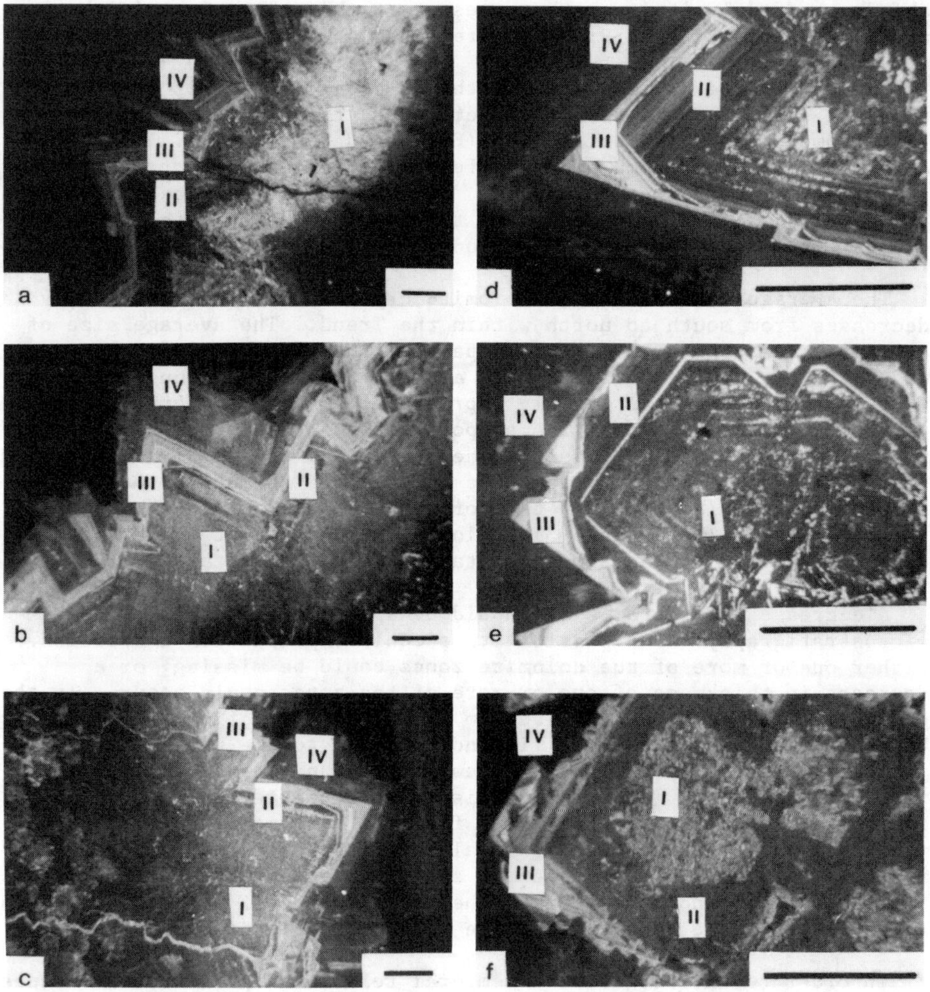

Figure 4. Cathodoluminescence photomicrograph showing dolomite rhombs from the following mines: a - Milliken, b - Fletcher, c - Brushy Creek, d - Buick, e - Magmont, and f - Viburnum No. 29. Note that the same four zones of dolomite spar can be distinguished at each mine. Scale bar = 250 micrometers.

more than a dozen headings. An exception to this systematic variation occurs at the Magmont mine where there is a slight increase in abundance of dolomite over that present at the other mines in the northern portion of the Viburnum Trend.

The thicknesses of each zone within a given crystal is dependent upon the size of the enclosing fracture and the quantity of fluids that flowed through that fracture (Ebers and Kopp, 1979). Therefore, absolute thicknesses are not a useful parameter in the correlation of individual zones between crystals. Within a single crystal, however, the thickness of one zone compared to the other zones within that crystal is useful in its recognition.

The average size of gangue dolomite crystals also systematically decreases from south to north within the Trend. The average size of dolomite crystals in the southern part of the Trend is 3-5 mm across, whereas those in the northern part average one mm. Although crystals from any given mine can locally approach 10 mm in size, crystals of this size are not common. The compositions of the larger crystals show a corresponding increase in the percentage of zone IV. Large saddle-shaped dolomite crystals are composed of as much as 90% of zone IV with the remaining 10% in the form of a seed crystal composed of zones I through III (Figure 3). Dolomite crystals that exhibit well-developed saddle shapes always contain a large percentage of zone IV.

A decrease in crystal size should be apparent in the dolomite microstratigraphy. Two possibilities could explain such a decrease. Either one or more of the dolomite zones could be missing, or a decrease in thickness of one or more of the zones could occur. In the dolomite crystals from the Viburnum Trend, a combination of both cases has been observed. Zones I, II, and III thin to the north. Zone III is depleted to a width of only a few microns and may consist only of two or three bands. Zone II diminishes to a point where it can be recognized in only a few specimens from the Viburnum 29 mine. Zone I thickness is variable, but a general thinning has occurred to the north. Zone IV thickness taken as a percentage of total thickness increases to the north, although the total volume of zone IV decreases as a result of the general decrease in sparry dolomite.

In order to investigate the temporal relationship of dolomite zones as time correlatable events, individual sets of bands were identified and compared between samples collected from widely spaced locations. Figures 3b, 3c, and 3d show the results of this comparison in cathodoluminescence photomicrographs of dolomite rhombs collected at various locations along the Trend, representing a distance of 29 kilometers. It is evident from an examination of these photomicrographs that the same pattern of zones is observed in each of the specimens.

The close association of the ore minerals with the dolomite gangue suggests that the observed variations in dolomite may bear upon the source for the fluids that deposited the ores in the Viburnum Trend. In this respect, it has been found that the dolomite growth zones and the sparry dolomite content decrease from south to north within the trend. The deceases in dolomite crystal size and abundance to the north in the Trend are related to increased dissolution or non-deposition of sparry dolomite. If the abundance of dolomite and their growth zones can be expected to be greatest in those mines that are closest to the source of the fluids, then this would suggest that the source of the fluids that deposited the dolomite and probably the ores was to the south of the Viburnum Trend. Fluids entering the Trend from the south would be expected to become progressively depleted in magnesium as they deposited dolomite and moved northward where they would deposit lesser amounts of dolomite. The fact that there are local deviations from this general systematic pattern, such as those observed at the Magmont, suggest that multiple entry points also were involved.

ELECTRON MICROPROBE ANALYSES

The major and minor element composition for each dolomite zone has been determined by electron microprobe analyses in order to investigate the chemical variations between dolomite zones and to determine if the composition within a given zone is consistent throughout the Trend. Ten representative samples were selected for the analyses, two each from the Milliken, Fletcher, Brushy Creek, Magmont, and Viburnum 29 mines. The samples were selected from similar stratigraphic horizons and at equal spacings along the strike of the Trend. Two to three crystals in each polished thin section were analyzed using a spot diameter of 20-30 micrometers. Each zone was analyzed for calcium, magnesium, iron, and manganese.

The results of the microprobe analyses indicate significant changes in the chemical environment during the deposition of the four major zones. Table I shows that the fluids that deposited zones I and II had iron and manganese concentrations that were moderate compared to those that deposited zone III. Zone III is very low in iron and mangangese. Zone IV has the largest iron content but similar manganese to that of zone III. These data suggest that the four dolomite zones were deposited under diverse chemical conditions.

A certain degree of scatter in the results was caused by the relatively large size of the spot diameter, the degree of internal banding within a zone, and the local presence of mottled textures. The microprobe backscatter images revealed subzones and additional internal banding not shown by cathodoluminescence microscopy.

Table 1. Average weight percent of CaO, MgO, FeO, and MnO for the four major dolomite growth zones.

Dolomite Zones		Average Weight Percent* CaO	MgO	FeO	MnO	FeO/MnO	CO_2	Total
Zone I		30.47	20.18	1.62	0.31	5.23	47.31	100.04
	SD =	0.39	0.58	0.62	0.13		0.54	1.02
Zone II		30.35	20.03	1.86	0.35	5.31	47.15	99.73
	SD =	0.90	0.60	0.55	0.09		0.41	0.76
Zone III		31.08	21.06	0.38	0.12	3.17	47.81	100.59
	SD =	0.66	0.56	0.25	0.13		0.54	0.86
Zone IV		31.07	19.45	2.02	0.13	15.54	47.11	99.89
	SD =	0.42	0.93	0.99	0.06		0.57	0.94

* Includes values for subzones
SD = Standard deviation

PARAGENETIC RELATIONSHIPS BETWEEN DOLOMITE AND SULFIDE MINERALS

Although the determination of paragenetic sequences in Mississippi Valley-type ore deposits is facilitated by the coarse grain size of the constitutent minerals and by simple superposition textures developed in open spaces, those studies are found to be surprisingly difficult due to other factors. The most important of these factors is that of repeated deposition of both the sulfide minerals and the gangue minerals, together with overlap in their periods of deposition. It is difficult to select a mineral as a useful relative time marker. Replacement and dissolution further complicate the interpretation. Some of the textures that are especially difficult to interpret can be resolved, however, with the use of plane polarized light petrography combined with cathodoluminescence microscopy.

By these combined techniques it was possible to determine that sphalerite had replaced host dolomite and dolomite of zone I in several samples (Figure 5a). Co-precipitation of sulfides and dolomite gangue are rarely observed under reflected light microscopy, and Anderson (1983) has suggested that the precipitation of sulfides would significantly limit the co-precipitation of dolomite spar.

Excellent cubes of galena have been found to occur at the base of sparry dolomite crystals that can be seen by cathodoluminescence to contain all four zones (Figure 5b). Two interpretations of this texture are possible. The galena may have been deposited in open space and subsequently was overgown by all four zones of dolomite. This interpretation would place the galena very early in the paragenetic sequence. Alternatively, the galena may be a metacryst that replaced the dolomite of zone I and therefore was deposited after dolomite. Unfortunately, there are no replacement remnants of dolomite within the galena cube which, if present, would attest to the latter interpretation.

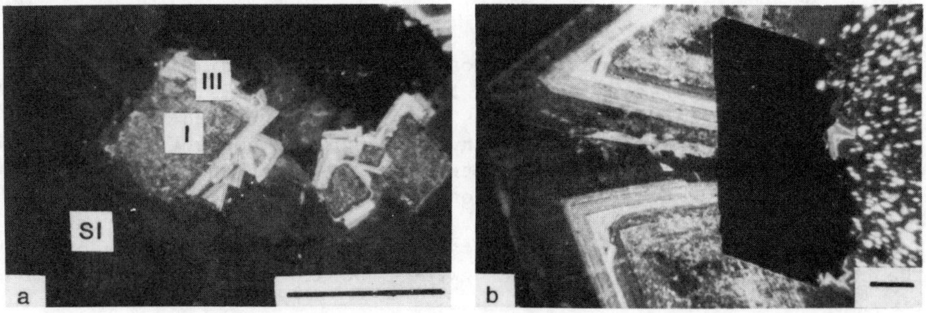

Figure 5. a - Cathodoluminescence photomicrography showing sphalerite (S1) filling open space and replacing zone I with later zone III filling remaining open space. b - Growth of spar on galena cube or replacement of spar by galena. Note the small dolomite crystal truncating against the base of the crystal; bright spots are potassium feldspar. Scale bar = 250 micrometers.

SUMMARY AND CONCLUSIONS

Cathodoluminescence microscopic study of 100 polished thin sections from eight mines in the Viburnum Trend subdistrict of the Southeast Missouri Lead District has shown that four growth zones can be consistently identified in sparry dolomite by distinctive cathodoluminescent characteristics. These four zones can be traced throughout a single mine, across stratigraphic units in the Bonneterre Formation, and through various types of structures, such as solution collapse breccias and slide breccias. Most importantly, the four zones can be traced from one mine to another mine throughout the entire length of the Viburnum Trend, a feature that strongly suggests that the fluids responsible for dolomite deposition were pervasive throughout the entire Trend.

Support for the interpretation that the cathodoluminescent zones are correlative comes from the fact that electron microprobe analyses have shown similar minor element chemistry for each of those zones from one mine to another. Electron microprobe results showed three periods of dolomite deposition, a high iron and manganese period (zone I and II), a low iron and manganese period (zone III), and a very high iron, low manganese period (zone IV). The low iron and manganese period, zone III, has been tied to sulfide deposition by paragenetic relationships. The zones I, II, and IV are associated with only minor periods of sulfide deposition.

In addition, even the fine internal cathodoluminescent banding can be traced for a distance of as much as 29 kilometers, a feature that further supports the idea of continuity of fluid movements throughout the trend.

The fact that four zones of deposition have been distinguished in sparry dolomite crystals indicates that those crystals were deposited from four distinct pulses of fluids that were introduced into the Viburnum Trend. The different iron and manganese contents for each of the main four dolomite zones as determined by electron probe analyses indicate that the chemical conditions were somewhat different during each period of dolomite deposition. The pesence of fine internal banding within the main bands indicates some degree of flucuation in the chemical conditions even during the deposition of a single band.

Most importantly, the four main zones of dolomite deposition can be traced from one mine to another mine throughout the entire length of the Viburnum Trend, a feature that strongly suggests that the fluids responsible for dolomite deposition were pervasive throughout the entire Trend.

ACKNOWLEDGEMENTS

This paper is the result of research partially funded under the auspices of the University of Missouri-Rolla. The authors are deeply indebted to the geology staffs at the operating mines. These include Bill Clendenin formerly at the Milliken mine, Ozark Lead Company; Milt Bradley and Laura Nadelhoffer at the Magmont mine, Cominco American Inc. and Dresser Industries Inc.; Randall Flanery at the Fletcher, Norm Paarlberg at the Brushy Creek, and Jim Husman and Bob Dunn at the Viburnum mines, St. Joe Minerals Corporation; Dave Burney at the Buick mine, AMAX Inc. and Homestake Mining Company. Without the help and expertise of the staff geologists, this research would not have been possible. A special thanks is expressed to Jay Gregg and the St. Joe Minerals Research laboratory for providing the electron microprobe analyses.

REFERENCES

Anderson, G. M., 1983, "Some Geochemical Aspects of Sulfide Precipitation in Carbonate Rocks", International Conference on Mississippi Valley Type Lead-Zinc Deposits, G. Kisvarsanyi, S. K. Grant, W. P. Pratt, and J. W. Koenig, eds., University of Missouri-Rolla, Rolla, MO, pp. 61-76.

Braunsdorf, N. R., and Lohman, K. C., 1983, "Isotopic Trends in Gangue Carbonates from the Viburnum Trend, Ozark Lead Mine, S.E. Missouri", Abstracts with Programs, 96th Annual Meeting, Geological Society of America, Indianapolis, Indiana, p. 532.

Ebers, M. L., and Kopp, O. C., 1979, "Cathodoluminescent Microstratigraphy in Gangue Dolomite, the Mascot-Jefferson City District, Tennessee", Economic Geology, Vol. 74, pp. 908-918.

Gerdemann, P. E., and Myers, H. E., 1972, "Relationships of Carbonate Facies Patterns to Ore Distribution and to Ore Genesis in the Southeast Missouri Lead District", Economic Geology, Vol. 67, pp. 426-433.

Hagni, R. D., 1983, "Ore Microscopy, Paragenetic Sequence, Trace Element Content, and Fluid Inclusions Studies of the Copper-Lead-Zinc Deposits of the Southeast Missouri Lead District", International Conference on Mississippi Valley Type Lead-Zinc Deposits, G. Kisvarsanyi, S. K. Grant, W. P. Pratt, and J. W. Koenig, eds., University of Missouri-Rolla, Rolla, MO, pp. 243-256.

Hagni, R. D., and Trancynger, T. C., 1977, "Sequence of Deposition of the Ore Minerals at the Magmont Mine, Viburnum Trend, Southeast Missouri", Economic Geology, Vol. 72, pp. 451-464.

Horrall, K. B., Hagni, R. D., and Kisvarsanyi, G., 1983, "Mineralogic, Textural and Paragenetic Studies of Selected Ore Deposits of the Southeast Missouri Lead-Zinc-Copper District and Their Genetic Implications", International Conference on Mississippi Valley Type Lead-Zinc Deposits, G. Kisvarsanyi, S. K. Grant, W. P. Pratt, and J. W. Koenig, eds., University of Missouri-Rolla, Rolla, MO, pp. 243-256.

Howe, W. B., 1968, "Planar Stromatolite and Burrowed Carbonate Mud Facies in Cambrian Strata of the St. Francois Mountain Area, Missouri Geological Survey Water Resources Report of Investigations 41, 113 pp.

Jessey, D. R., 1981, "An Investigation of the Nickel-Cobalt Occurrence in the Southeast Missouri Mining District", Unpublished Ph.D. dissertation, University of Missouri-Rolla, Rolla, MO, 225 pp.

Larsen, K. G., 1977, "Sedimentology of the Bonneterre Formation, Southeast Missouri", Economic Geology, Vol. 72, pp. 408-419.

Lyle, J. R., 1977, "Petrography and Carbonate Diagenesis of the Bonneterre Formation in the Viburnum Trend Area, Southeast Missouri", Economic Geology, Vol. 72, pp. 420-434.

Mugel, D. N., and Hagni, R. D., "Mineralogy and Geology of the Blanket Lead-Zinc Deposit, Buick Mine, Viburnum Trend, Southeast Missouri", ICAM 84: Proceedings of the International Congress on Applied Mineralogy in the Mineral Industry 84, Los Angeles, California, in press.

McLimans, R. K., Barnes, H. L., and Ohmoto, H., 1980, "Sphalerite Stratigraphy of the Upper Mississippi Valley Zinc-Lead District, Southwest Wisconsin", Economic Geology, Vol. 75, pp. 351-361.

Rickman, D. L., 1981, "A Thermochemical Study of the Ore Deposits of the Milliken Mine, New Lead Belt, Missouri", Unpublished Ph.D. dissertation, University of Missouri-Rolla, Rolla, MO, 295 pp.

Rowan, Lanier, 1984, Private communication.

Walkden, G. M., and Berry, J. R., 1984, "Natural Calcite in Cathodoluminescence: Crystal Growth during Diagenesis", Nature, Vol. 308, pp. 525-527.

URANIUM INVESTIGATIONS

Chapter 6

URANIUM, RARE EARTH AND IRON MINERALIZATION IN PEGMATITE AT THE BEMCO MINE PROSPECT, CRANBERRY LAKE, NEW JERSEY

Andreas H. Vassiliou and John H. Puffer

Department of Geological Sciences
Rutgers University
Newark, New Jersey 07102

ABSTRACT

The Bemco mine area is within the Precambrian igneous and metamorphic rock terrain of the New Jersey Highlands province. Rocks in the area include a pyroxene granite, a hornblende granite, a quartz-oligoclase gneiss and a pegmatite and its host rock, a lens of amphibolite which becomes a sulfide-rich pyroxenite near the ore zone. The ore-bearing zone occurs along the hanging wall and foot-wall contact of the pegmatite which was emplaced discordantly into the amphibolite-pyroxenite.

A 7 cm thick ore zone, in contact with the pegmatite, is composed of a fine-grained matrix of magnetite (up to 76%), is rich in metamict zircon crystals (up to 19%) and contains uranothorite (up to 9%) and fergusonite (up to 4%). A 25 cm thick zone, in contact with the pyroxenite, is essentially a magnetite ore (up to 92%) but it also contains accessory fergusonite and zircon (up to 3% and 4% respectively).

There are at least three plausible sources for the aqueous fluids that formed the ore-bearing pegmatite: metamorphic dehydration of the amphibolite; metamorphic dehydration of the quartz-oligoclase gneiss; late stage aqueous fluids from the granite magmas.

INTRODUCTION

The Bemco mine prospect is located just south-east of Cranberry Lake in the Stanhope quadrangle of northwestern New Jersey, about 15 miles south-east of the Franklin deposit.

Exploratory mining in 1959, mainly in the form of open cuts, revealed a narrow ore band (average width about 30 cm) along the strike of the contact between a granite pegmatite and the surrounding wall rock, a pyroxenite. Additional mining operations conducted primarily in 1959 included two vertical shafts and two adits driven several meters in a N-W direction at two different levels along the strike of the pegmatite.

Between 1959 and 1968 a total of 95 tons of ore was sold for its uranium content, but its high rare-earth values were largely overlooked. The ore consisted of a friable matrix of magnetite (about 90%) and a suite of accessory ore minerals such as zircon, fergusonite and uranothorite. The property is owned by the State Parks development but mineral and mining rights have been retained by the previous owner who has prospected the mine area as recently as 1978 (Tosch, 1981).

Diamond drill cores studied by Williams (1967) showed that the pegmatite as well as the ore zone narrow considerably with depth; however, the ore was shown to be quite rich in rare-earth oxides (up to 2.4%) and to contain 0.45% U_3O_8 and 0.24% ThO_2. Vassiliou, Puffer and Markewicz (1974) and Vassiliou (1980; 1980a) described the mineral assemblage in the ore. Vassiliou (1980a) suggested that the ore was probably precipitated from hydrothermal solutions associated with the late stages of the intrusion of the pegmatite.

This report describes the geologic setting of the mine area and analyzes the mineralogy of the ore and the associated rock units in light of the latest information on the petrogenesis of the region.

GEOLOGIC SETTING

The Bemco mine area is located within the Precambrian igneous and granite facies metamorphic rock terrain of the New Jersey Highlands province which is part of what Drake (1984) describes as the Proterozoic Y terrain of the Reading Prong of eastern Pennsylvania and New Jersey. The mine lies near the central portion of what Young (1978) refers to as the Sparta Mountain-Cranberry Lake Block. The N.E. trending fault block is bounded on the N.W. by the zero fault zone (Baley, Salisbury and Kummel, 1914) and on the S.E. by the East fault zone (Hague et al., 1956). The block is split in half by the N-E trending Wright Pond Fault that cuts through the map area (Fig. 1). The downthrown side is on the S-E side of the Wright Pond Fault. The pegmatite is located within the downthrown side. The foliation of these rocks strikes N 60°W and dips at a high angle (30° to 80°) to the north. The Precambrian rocks that occur at the mine area (Fig. 1) include a Pyroxene Granite and a Hornblende Granite which are part of Drake's (1984) Byram Intrusive Suite, a Quartz-Oligoclase Gneiss and an Amphibolite-Pyroxenite which are part of Drake's (1984) Losee Metamorphic Suite and a Pegmatite which was emplaced discordantly into the Amphibolite-Pyroxenite.

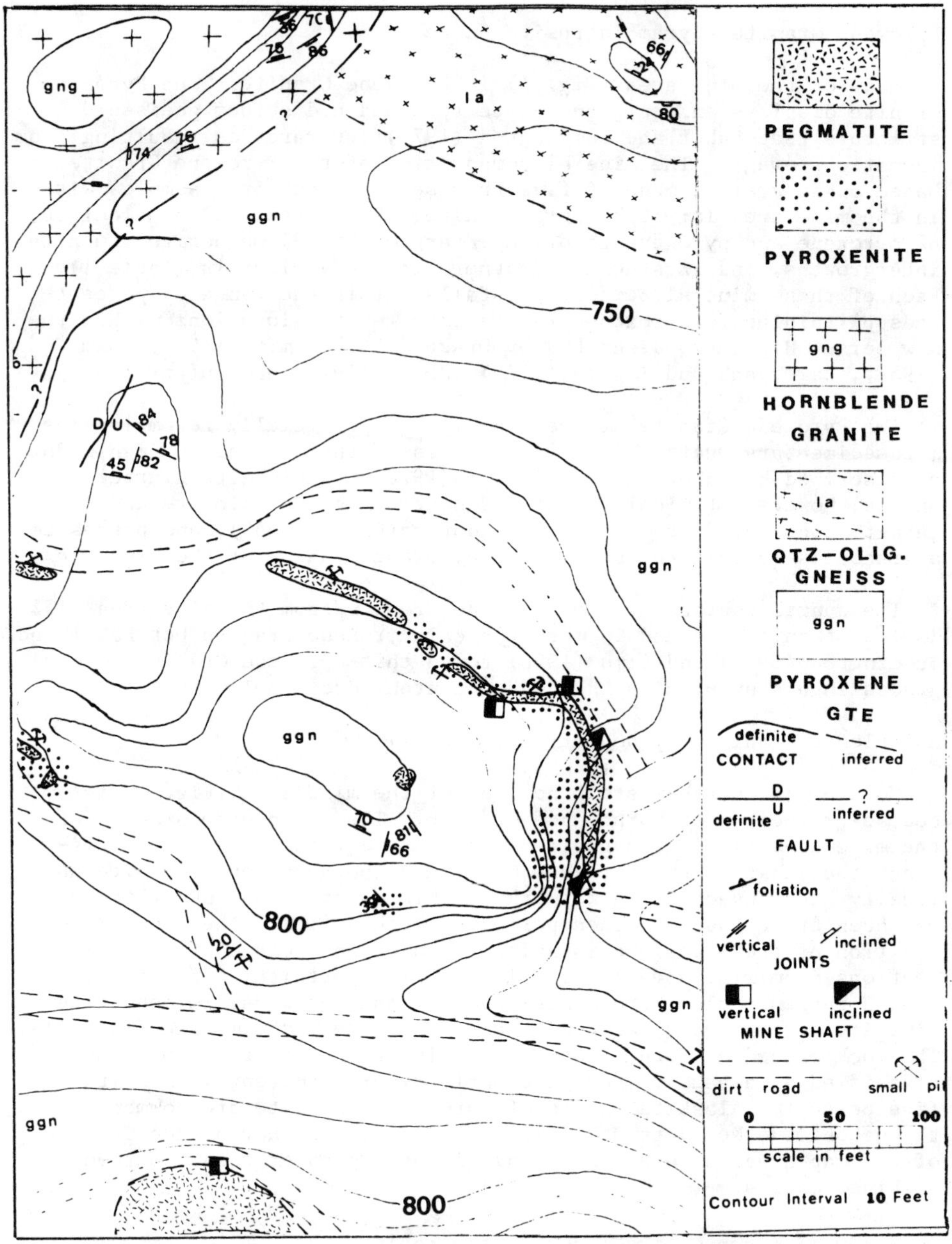

Fig. 1. Geologic map of the Bemco Mine prospect. Field data provided by D. Hurtubise, W. Orlowski and D. Fontaine.

Pyroxene Granite (Byram Intrusive Suite)

Most of the mine area (Fig. 1) is Pyroxene Granite. The Pyroxene Granite displays an equigranular texture and a distinct gneissoid structure that Baker and Buddington (1970) interpret as resulting from magmatic flowage. The mineral composition of the Pyroxene Granite based on an average mode of five thin sections cut from samples within the mine area is: 65% microperthite, 15% quartz, 10% plagioclase, 6% pyroxene and pyroxene-chlorite intergrowths, 3% magnetite-ilmenite intergrowths, and 1% sphene and other accessory minerals (Table 1). Each of these mineral percentages falls within the range of percentages pertaining to pyroxene granite from several locations within the New Jersey Highlands described by Drake (1969), Smith (1969), Sims (1958), Baker and Buddington (1970), Rhett (1977) and Puffer (1980).

The Pyroxene Granite at the mine area superficially resembles the metasedimentary Quartz-Microcline Gneiss of the New Jersey Highlands as described by Baker and Buddington (1970). The Quartz-Microcline Gneiss, however, typically contains several percent biotite and/or garnet, contains 36 to 72 percent quartz (Drake, 1969), and possesses a granoblastic texture, unlike the Pyroxene Granite of the mine area.

The quartz content of the Pyroxene Granite from the mine area (15%) is less than that found in more typical Pyroxene Granite but Baker and Buddington (1970) and Sims (1958) found that Pyroxene Granite commonly grades into a syenite as the quartz content drops below 10 percent.

Hornblende Granite (Byram Intrusive Suite)

Most of the north-western portion of the mine area (Fig. 1) is coarse-grained, pink Hornblende Granite. The Hornblende Granite of the mine area displays a gneissoid structure that is much less distinct than that of the Pyroxene Granite. The Hornblende Granite has locally been invaded by a network of thin quartz veins as if the rock had been brecciated and then permeated with silica. The mineral composition of the Hornblende Granite based on an average of six thin sections from mine area samples is 55% microperthite, 34% quartz, 4.5% hornblende plus hornblende-chlorite intergrowths, 1% magnetite-ilmenite intergrowths, with a trace of apatite and ilmenite (Table 1). The rock mapped as Hornblende Granite includes some rock that may be classified as alaskite where the mafic mineral content is less than five percent. Alaskite facies of Hornblende Granite are common throughout the New Jersey Highlands. Most of the Hornblende Granite of the map area, however, contains slightly greater than 5 percent combined mafics.

Quartz-Oligoclase Gneiss (Losee Metamorphic Suite)

The north-eastern portion of the mine area (Fig. 1) and a small area south of the mine is composed of Quartz-Oligoclase Gneiss. The rock appears white, medium grained, homogeneous, and foliated. The

TABLE 1. Modal Composition of Rock Types at the Bemco Mine Prospect

Mineral	Pyroxene Granite	Hornblende Granite	Quartz-Olig. Gneiss	Amphibolite	Pyroxenite*		Pegmatite	
					PO	PM	Core	Margin/ore
K-feldspar[1]	65	55	4		43	8	56	10
Plagioclase[1]	10		60	48	8	12	10	28
Quartz	15	34	34	3	2		31	20
Pyroxene[2]	6			12	40	65	tr	6
Hornblende[2]		4.5	tr	30	2	9	tr	tr
Biotite			1	3	2	tr		2
Magnetite[3]	3	1	1	2	2	tr	tr	32
Ilmenite	tr	tr		1		tr		tr
Pyrite	tr			2	tr	2		tr
Pyrrhotite				1	tr	2		
Chalcopyrite				tr		1		
Zircon	tr			tr				tr
Sphene	1			2				
Apatite	tr	tr	tr	tr				tr
Fluorite								tr
Allanite								tr
Fergusonite								tr

* PO = near ore zone contact
PM = approx. one meter from ore contact

1. Plus perthite or antiperthite
2. Plus chlorite intergrowths
3. Plus ilmenite intergrowths

mineral composition of the gneiss based on an average of five thin sections from mine area samples is: 60% oligoclase, 34% quartz, 4% microcline, 1% biotite, 1% magnetite-ilmenite intergrowths, with traces of amphibolite and apatite (Table 1). The Quartz-Oligoclase Gneiss contains several small magnetite-bearing granite pegmatites that are emplaced concordant to the foliation of the gneiss.

Amphibolite-Pyroxenite (Losee Metomorphic Suite)

A lens of amphibolite is the host rock for the rare-earth bearing pegmatite (Fig. 1). Numerous amphibolite bands and lenses as small as a few centimeters across occur within each of the major igneous and metamorphic rock units of the map area but are distributed too irregularly to be mappable. The mineral composition of the amphibolite is highly variable throughout the map area but on the basis of an average mode of six thin sections the rock is composed of 45% plagioclase, 30% hornblende, 12% pyroxene, 3% sulfides, 3% quartz, 3% biotite, 2% sphene, 2% magnetite-ilmenite intergrowths, 1% ilmenite, with traces of apatite and zircon. Except for the unusually high sulfide content of the amphibolite, the rock is rather typical of the amphibolites of the New Jersey Highlands as described by Maxey (1971). On the basis of major and trace element chemical analyses of fifty-six samples of amphibolites from locations scattered throughout the New Jersey Highlands, Maxey (1971) concluded that the amphibolites are metamorphosed tholeiitic basalts. Maxey (1971) also described highly variable pyroxene/hornblende ratios throughout the New Jersey Highlands.

The pyroxene and sulfide contents of the amphibolite within the map area (Fig. 1) are particularly variable. As the ore-bearing zone of the pegmatite is approached the amphibolite becomes a very coarse-grained and sulfide-rich pyroxenite. The pyroxene content increases to 65%, the K-feldspar content increases to 29% and the sulfide content increases to 7% adjacent to the ore deposit while the plagioclase content decreases to 18% and the hornblende content decreases to 2%. The majority of the sulfide content is pyrrhotite while most of the remaining sulfide is pyrite. Chalcopyrite, is also commonly found in the amphibolite with minor amounts of galena and sphalerite.

Pegmatite

Although numerous small, concordant and unzoned pegmatites are found in each of the major rock units of the map area (Fig. 1), only one large (approx. 2 m thick), zoned and ore-bearing pegmatite has been found. The ore-bearing pegmatite has an approximately E-W strike and dips to the south at an angle of approximately 60°. The strike length of the pegmatite is approximately 265 m. Diamond drilling data indicates that at 33 m depth the thickness of the pegmatite has been slightly displaced by several N-S trending faults described in detail by Van de Kamp (1962). The pegmatite has been discordantly intruded into Amphibolite-Pyroxenite.

A gradational, indistinct, zoning can be identified within the pegmatite upon careful observation (Table 1). The core of the pegmatite is much richer in perthite (56%) and quartz (31%) than the outer zone containing 10% perthite and 20% quartz. The core also contains less magnetite (tr), plagioclase (10%), pyroxene (tr) and biotite (tr) than the outer zone that contains 32% magnetite, 28% plagioclase, 6% pyroxene and 2% biotite. Accessory hornblende and trace amounts of ilmenite, pyrite, zircon, fergusonite, allanite, apatite, and fluorite are also found in the pegmatite. The ore-bearing zone occurs along both the hanging wall and footwall contact of the pegmatite.

MINERALOGY

The ore minerals occur almost exclusively in narrow zones or bands along the hanging and footwall contacts between the pegmatite and the pyroxenite. The zones average approximately 30 cm in width on both walls but there is considerable pinching and swelling and in some cases the zones are absent. Diamond drill cores show that the ore zone narrows considerably with depth; at a depth of about 30 meters, the zone has a maximum width of about 2.5 cm (Williams, 1967).

Megascopically as well as optically the ore can be divided into two types. The spatial relationship between the two types is shown in Fig. 2. Type "A" ore represents a 7 cm thick zone, in contact with the pegmatite, that may be described as rusty or oxidized. It is composed of a fine-grained matrix of magnetite (up to 76%) that is rich in zircon in the form of euhedral crystals disseminated throughout the matrix. It is also relatively rich in uranothorite (up to 9%) and fergusonite (up to 4%). Type "B" ore represents a 25 cm thick zone, in contact with the pyroxenite, that is essentially a magnetite ore (up to 92% magnetite), but it also contains accessory fergusonite (up to 3%) and zircon (up to 4%). Fergusonite and zircon are not readily recognized magascopically in this ore. The modal composition for both types of ore is shown in Table 2.

The ore minerals zircon, fergusonite and uranothorite are radioactive and metamict. Their oxide composition, representing quantitative estimations obtained through X-ray fluorescence analysis using artificially prepared standards, is shown in Table 3. A few black anhedral grains separated from the ore have been tentatively identified (through X-ray powder camera) as the yttrium silicate "rowlandite" by Fontaine (1976). The relatively rare metamict mineral "spencite"(calcium-yttrium silicate) has been reported as present in the ore by Williams (1967), but its presence has not been confirmed by the writers. A more detailed description of the ore minerals follows.

Magnetite

As much as 90% of the ore zone or contact zone is composed of coarse-grained magnetite. In addition, magnetite occurs in the form

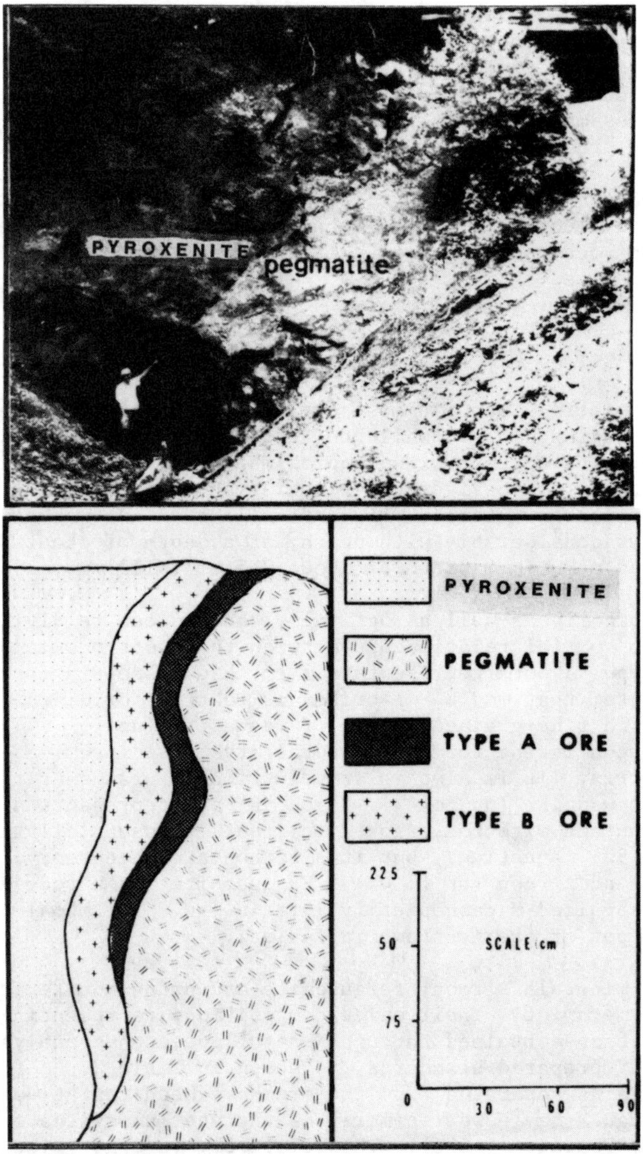

Fig. 2. (Photo) The pegmatite-pyroxenite contact at the Bemco Mine works.
(Diagram) Schematic cross-section illustrating the spatial relationships between the associated rock units and the ore at their contact.

TABLE 2. Modal Composition of Ore at Bemco Mine

Mineral	Type "A" Near Pegmatite Contact		Type "B" Near Pyroxenite Contact	
Feldspar				tr
Quartz				tr
Pyroxene			2	1
Hornblende		43	1	1
Magnetite	76	14	92	91
Sulfides		tr	tr	tr
Zircon	14	19	1	4
Fergusonite	2	4	3	2
Uranothorite	1	9	tr	tr
Allanite		tr		tr
Apatite	7	8		tr
Fluorite				tr

TABLE 3. Composition of Ore Minerals at Bemco Mine

	Fergusonite	Uranothorite	Zircon
ThO_2	10.2	50.3	6.7
UO_3*	4.3	26.2	2.8
ZrO_2			59.3
Nb_2O_3	36.4		
Y_2O_3	27.9	0.7	0.3
Fe_2O_3*	4.7	5.1	1.1
Yb_2O_3	1.7		2.2
Ta_2O_5	2.9		
Er_2O_3	2.1		
Gd_2O_3	1.1		
HfO_2			0.8
CeO_2*		2.1	
As_2O_3		1.0	0.5
Total	91.3	85.4	73.7

*Oxidation state unknown

of veinlets (up to 16 cm in length) which cut through the pegmatite as well as its host rock, the pyroxenite. It is also commonly disseminated in both the pegmatite and the pyroxenite as well as in all the other rocks in the Cranberry Lake area.

The magnetite of the ore zone and the related veinlets is identical in chemical composition to the magnetite found in the pegmatite, with a TiO_2 content of only 0.1%. On the other hand, the magnetite found disseminated in the pyroxenite host rock contains up to 3.5% TiO_2. This suggests a direct genetic link between the ore and the pegmatite.

Zircon

Euhedral crystals, mainly concentrated in the type "A" ore, average about 4 mm in length, with some as long as 2 cm. These crystals are generally gray, strongly zoned, and some exhibit twinning. The zoning is visible in most crystals observed in thin section, with zircon forming the crystal walls (cyrtolite), and uranothorite and fergusonite occupying the core (malacon) of the crystals (Fig. 3A).

In Table 3 the zircons ($ZrSiO_4$) are shown to be rich in thorium and uranium. The most abundant rare earth present is ytterbium, with other associated elements being hafnium and yttrium.

The crystals are completely metamict. However, with heat treatment (1000°C for one hour in a muffle furnace), the crystal structure of zircon could be ascertained through X-ray diffraction analysis.

Fergusonite

This mineral occurs primarily in the Type "B" ore, especially as part of the core in zoned zircons. Optically it appears as red-brown vitreous grains that often contain magnetite inclusions (Fig. 3B).

Table 2 shows fergusonite, $Y(Nb, Ta)O_4$, to be rich in thorium, uranium, and several rare earths (ytterbium, erbium, gadolinium) in addition to elements listed in its chemical formula.

The mineral is metamict and, depending on heat treatment, it may exhibit one of several phases. Lima de Faria (1964, Table 18) lists four phases. In this study, the mineral was heated (in a muffle furnace) at 1000°C for one hour, and its diffraction pattern correlated well with the monoclinic phase outlined by Lima de Faria (1964, Table 20).

Uranothorite

This mineral occurs mainly in the cores of zoned zircons (Fig. 3A) and occasionally in the form of veinlets in the ore matrix. Optically it appears as black, isotropic subhedral grains (medium-to-coarse-grained), and its identification was confirmed by X-ray diffraction

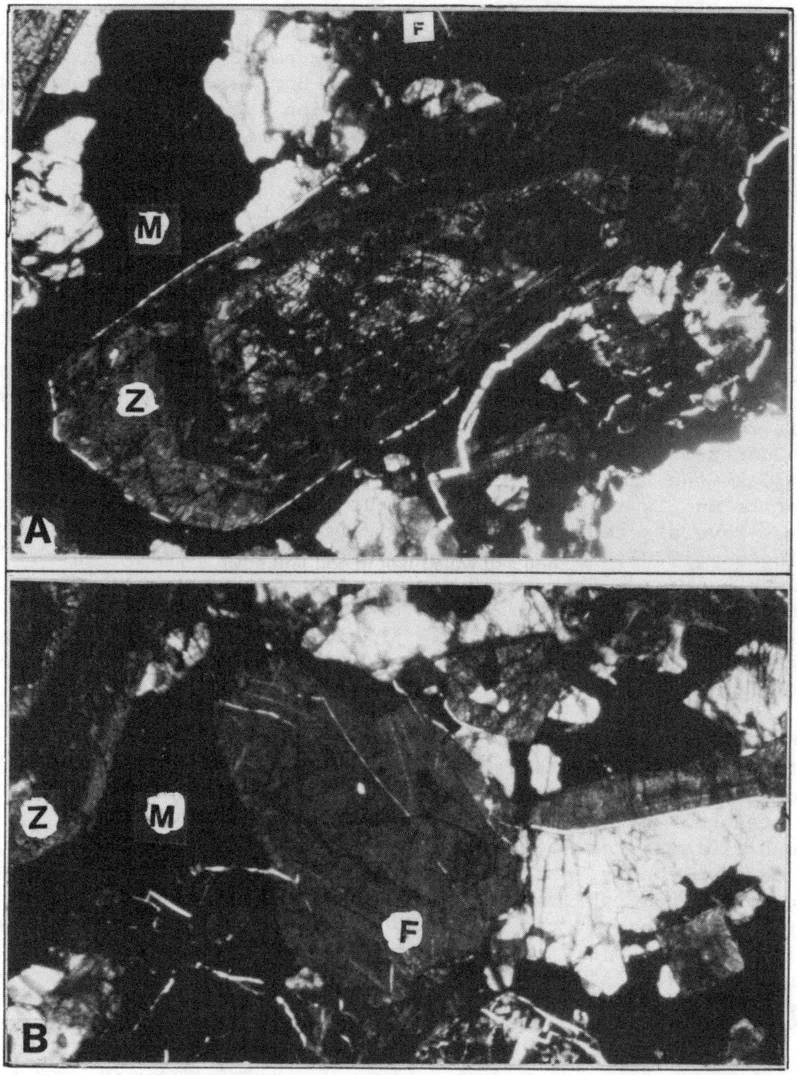

Fig. 3. A. Zoned zircon crystal (Z) with associated magnetite (M) and fergusonite (F) in type "A" ore. The outer zone of the crystal is cyrtolite and the core is malacon which contains uranothorite and some fergusonite.

B. Large fergusonite grain (F) in association with magnetite (M) and zircon (Z) in type "A" ore.

Plain polarized light, 16X.
Photomicrographs by L.M. Fukui and R.F. D'Andrea, Jr.

after heating at 1000° for one hour. Its X-ray diffraction pattern is very close to the monoclinic huttonite phase of thorite ($ThSiO_4$) as shown by Lima de Faria (1964, Table 1). However, the presence of large amounts of uranium in its composition (see Table 3) suggests uranothorite, (Th, U) SiO_4, as a more appropriate name.

PETROGENESIS

In Northern New Jersey there seems to be a genetic relationship between uraniferous rare-earth deposits, magnetite deposits and pegmatites. Williams (1967) found rare earth mineral concentrations in 9 out of 11 iron mines investigated within the Northern New Jersey area. High levels of rare-earth concentrations approaching those found at the Bemco prospect were found at the Scrub Oaks iron mine of Morris County, New Jersey. Most of these iron mines were emplaced into Quartz-Oligoclase Gneiss. Klemic et al. (1959) found that at Scrub Oaks the rare earth ore is concentrated in coarse-grained magnetite and is in pegmatite adjacent to the magnetite concentrations. Several studies including Klemic et al. (1959), Baker and Buddington (1970), Sims (1958) and Puffer (1980) have shown that although pegmatites are found distributed throughout each of the pre-cambrian rock types of the New Jersey Highlands they are most abundant within magnetite concentrations. K-feldspar rich pegmatites are an important part of each of the magnetite deposits studied by Puffer (1980).

The rare-earth concentrations, the magnetite concentrations, and at least some of the Highlands pegmatites, therefore, probably have a common source. Most experimental evidence (Jahns and Burnham, 1969) indicates that pegmatites precipitate out of aqueous fluids. The same aqueous fluid system may have also transported considerable iron and rare earth elements. Magnetite rich pegmatites are commonly found throughout metamorphic and igneous terrains throughout North America (Puffer 1975), and it can be demonstrated experimentally that iron is partitioned into fluid phases separated from silicate melts (Martin and Piwinski 1969). At the Bemco Mine area the iron apparently precipitated as magnetite at a late stage of pegmatite formation and occupies the outermost zone. The pegmatite fluids were undersaturated in titanium resulting in unusually low titanium concentrations in the ore zone magnetite (0.1%) and the absence of any coexisting titanium oxide phase within the pegmatite. The same aqueous fluid system probably also transported the zircon, uranium, and rare earths that are concentrated in the ore zone of the pegmatite. Each of these elements are "incompatable elements" and are typically concentrated in late stage igneous differentiates and associated pegmatites. Pegmatites are common host rock for uranium (Klepper and Wyant, 1957).

Textural evidence outlined under "Mineralogy" (see Fig. 3) indicates that the precipitation of the rare earths followed the precipitation of the magnetite and zircon. At still lower temperatures as the aqueous fluids diffused into the amphibolite-pyroxenite host rock more iron and some copper precipitated as sulfides.

There are at least three plausible sources of the aqueous fluids that formed the ore-bearing pegmatite: Metamorphic dehydration of the Amphibolite; metamorphic dehydration of the Quartz-Oligoclase Gneiss; and late stage igneous release of aqueous fluids from the Granite magmas.

Metamorphic Dehydration of the Amphibolite

Collins (1969) suggested that the metamorphic diffusion processes account for the magnetite concentration found in amphibolites within the Dover district of New Jersey. His evidence includes a pronounced decrease in the iron content of the amphibolite host rocks as magnetite concentrations are approached. He concludes that the iron removed from amphibolites through metamorphic diffusion is approximately equivalent to the volume of iron contained within the ore deposits.

Similar metamorphic processes may account for the iron concentration at the Bemco Mine. But unlike the concordant pegmatite and magnetite concentrations at most Dover district mines, the pegmatite at the Bemco Mine is discordant and appears to have been forcibly intruded into the amphibolite-pyroxenite from an outside source, at a high angle to the foliation of the host rock. In addition, the analysis of a series of samples of amphibolite-pyroxenite, taken at the ore contact and at regular intervals away from the contact, shows that the iron content (Fe_2O_3 and total Fe content) decreases away from the contact. This suggests that the magnetite ore is probably not related to metamorphic diffusion processes or to deuteric leaching of iron from the amphibolite-pyroxenite.

Metamorphic Dehydration of the Quartz-Oligoclase Gneiss

If the Quartz-Oligoclase Gneiss is a metasedimentary rock as suggested by Sims (1958), Collins (1969) and Puffer (1980), it presumably underwent considerable prograde metamorphism to reach the granulite facies. Since water is generally released during each progressive stage of metamorphism, a solvent is made available for the iron, potassium and rare earths concentrated in the pegmatite. Metamorphic reactions involving the breakdown of biotite may have released considerable iron, potassium, rare earths and water. These elements may have been transported as an aqueous fluid and emplaced into perhaps a shear zone within the amphibolite-pyroxenite. Precipitation of iron may have been forced by decreasing temperatures and/or pressures. Puffer (1980) suggested this metamorphic mechanism may be responsible for some of the iron deposits found within the New Jersey Highlands. Support for this suggestion includes:

(a) The low iron and potassium content of the Quartz-Oligoclase Gneiss. The chemical composition of the Quartz-Oligoclase Gneiss is approximately equivalent to either typical graywacke or tonalite. However, regardless of whether the precursor of the Gneiss was a graywacke (Puffer, 1980) or an intrusive igneous rock, presumably a tonalite (Baker and Buddington, 1970), the iron and potassium content is unusually low for either rock type. The chemical composition of the Quartz-Oligoclase Gneiss also resembles a quartz keratophyre, as pointed out by Drake (1967). The amphibolite bands found within the Quartz-Oligoclase Gneiss, however, have been determined by Maxey (1974) to be meta-tholeiitic basalt layers. Tholeiitic basalt and quartz keratophyre are rather incompatable volcanic rock and rarely, if ever, occur together in nature.

(b) The close spatial association of Quartz-Oligoclase Gneiss with the Bemco pegmatite and with most iron ore - pegmatite associations within the New Jersey Highlands.

Late Stage Aqueous Fluids from the Granite Magmas

The Bemco pegmatite may have been generated out of late stage magmatic processes associated with Granite intrusions. The pegmatite is discordant and may have been injected into the metamorphic host rock intruded by the granite. Baker and Buddington (1970) and Sims (1958) have suggested that hydrothermal fluids released by granitic intrusions are responsible for the emplacement of the magnetite concentrations of the New Jersey Highlands. The same granite magma may also have emplaced the Bemco pegmatite. Their suggestion is supported by arguments that are very similar to those supporting the Quartz-Oligoclase Gneiss source:

(a) The low iron content of an alaskite facies of granite that may have developed as iron was released from Hornblende Granite magma.

(b) The close spatial association of the Hornblende Granite alaskite with most iron ore - pegmatite associations within the New Jersey Highlands.

Conclusion

The above three sources of the aqueous fluids that formed the ore-bearing pegmatite are presented as plausible alternate hypotheses. Of the three, the metamorphic dehydration of the Amphobolite appears to be the least likely source for reasons outlined in the text. However, even if we eliminate this hypothesis, it may never be possible to determine which of the other two rock types, the Quartz Oligoclase Gneiss or the Granite magmas, was the source of the fluids that formed the pegmatite and the associated iron, uranium and rare-earth mineralization at the Bemco Mine prospect.

REFERENCES

Baker, D.R., and Buddington, A.F., 1970, "Geology and Magnetite Deposits of the Franklin Quadrangle and Part of the Hamberg Quadrangle, New Jersey", Professional Paper 638, U.S. Geological Survey, 73 pp.

Bayley, W.S., Salisbury, R.D., and Kummel, H.B., 1914, "Raritan, N. J.", Folio 191, U.S. Geological Survey Geological Atlas, 32 pp.

Collins, L.G., 1969, "Regional Recrystallization and the Formation of Magnetite Concentrations, Dover Magnetite District, New Jersey", Economic Geology, vol. 64, pp. 17-33.

Drake, A.A., Jr., 1967, "Geologic Map of the Easton Quadrangle, New Jersey-Pennsylvania", Geological Quadrangle Map GQ-594, U.S. Geological Survey.

Drake, A.A., Jr., 1969, "Precambrian and Lower Paleozoic Geology of the Delaware Valley, New Jersey-Pennsylvania", Geology of Selected Areas in New Jersey and Eastern Pennsylvania and Guidebook of Excursions, Subitzky, S., ed., Rutgers University Press, New Brunswick, N.J., pp. 51-131.

Drake, A.A., Jr., 1984, "The Reading Prong of New Jersey and Eastern Pennsylvania: An Appraisal of Rock Relations and Chemistry of a Major Proterozoic Terrane in the Appalachians", Geological Society of America Special Paper 194, pp. 75-109.

Fontaine, D.A., 1976, "The Geology and Ore Genesis of the Bemco Rare-Earth Deposit at Cranberry Lake, New Jersey," M. Sc. Thesis, Rutgers University, Newark, 40 pp.

Hague, J.M., et al., 1956, "Geology and Structure of the Franklin-Sterling area, New Jersey", The Geological Society of America Bulletin, vol. 67, pp. 435-474.

Jahns, R.H., and Burnham, W.C., 1969, "Experimental Studies of Pegmatite Genesis: I. A Model for the Derivation and Crystallization of Granite Pegmatites", Economic Geology, vol. 64, pp. 42-46.

Klemic, H., et al., 1959, "Radioactive Rare-Earth Deposits at Scrub Oaks Mine, Morris County, New Jersey", Bulletin 1082-B, U.S. Geological Survey, pp. 29-59.

Klepper, M.R., and Wyant, D.G., 1957, "Notes on the Geology of Uranium", Bulletin 1046-F, U.S. Geological Survey, 148 pp.

Lima de Faria, J., 1964, "Identification of Metamict Minerals by X-ray Powder Photographs", Junta de Investigacoes do Ultramar, No. 112,

Lisboa, Portugal, 74 pp.

Martin, R.F., and Piwinski, A.J., 1969, "Experimental Data Bearing on the Movement of Iron in an Aqueous Vapor", Economic Geology, vol. 64, pp. 798-803.

Maxey, L.R., 1971, "Metamorphism and Origin of Precambrian Amphibolite of the New Jersey Highlands", Ph.D. Dissertation, Rutgers University, New Brunswick, 156 pp.

Puffer, J.H., 1975, "Some North American Iron-Titanium Oxide Bearing Pegmatites", American Journal of Science, vol. 275, pp. 708-730.

Puffer, J.H., 1980, "Precambrian Rocks of the New Jersey Highlands", Field Studies of New Jersey Geology and Guide to Field Trips, Manspeizer, W., ed., Rutgers University, Newark, N.J., pp. 42-53.

Rhett, D.W., 1977, "Phase Relationships and Petrogenetic Environment of Precambrian Granites", Ph.D. Dissertation, Rutgers University, New Brunswick, 157 pp.

Sims, P.K., 1958, "Geology and Magnetite Deposits of the Dover District, Morris County, New Jersey", Professional Paper 287, U.S. Geological Survey, 162 pp.

Smith, B.L., 1969, "The Precambrian Geology of the Central and Northeastern Parts of the New Jersey Highlands", Geology of Selected Areas in New Jersey and Eastern Pennsylvania and Guidebook of Excursions, Subitzky, S., ed., Rutgers University Press, New Brunswick, N.J., pp. 35-47.

Tosch, K.W., 1981, "Investigation of the Charlotte Mine", Bureau of Radiation Protection Report, New Jersey State Department of Environmental Protection, 9 pp.

Van de Kamp, P.G., 1962, "Rare-Earth Mineral Deposit at Cranberry Lake, New Jersey", Senior Thesis, Lehigh University, 26 pp.

Vassiliou, A.H., 1980, "Metamict Minerals at the Bemco Mine Near Cranberry Lake, New Jersey", The Mineralogical Record, vol. 11, p. 39.

Vassiliou, A.H., 1980a, "Uranium and Rare Earth Mineralization at the Bemco Mine Near Cranberry Lake, New Jersey", Field Studies of New Jersey Geology and Guide to Field Trips, Manspeizer, W., ed., Rutgers University, Newark, N.J., pp. 192-199.

Vassiliou, A.H., Puffer, J.H., and Markewicz, F.J., 1974, "Uranium-Rare Earth Mineralization at Charlotte Mine Prospect Near Cranberry Lake, New Jersey", Geological Society of America Northeastern Section Abstracts, vol. 5, p. 33.

Williams, R.L., 1967, "Reconnaissance of Yttrium and Rare-Earth Resources in Northern New Jersey", U.S. Bureau of Mines Report of Investigations 6885, pp. 9-15.

Young, D.A., 1978, "Precambrian Salic Intrusive Rocks of the Reading Prong", The Geological Society of America Bulletin, vol. 89, pp. 1502-1514.

Chapter 7

MINERALOGY AND GEOLOGY OF TEXACO'S[1] HOBSON URANIUM DEPOSIT, KARNES COUNTY, TEXAS

by Harry M. Dahl and Christie A. Callender

Texaco Houston Research Center,
Bellaire Research Facility
Bellaire, Texas 77401

ABSTRACT

The Hobson uranium deposit is located in Karnes County, Texas, along the prominent south Texas uranium trend. The uranium mineralization occurs in the Deweesville Sandstone of the Whitsett Formation of the Eocene Jackson Group. The host rock is a friable, fine-grained, well-sorted, highly porous and permeable sandstone interpreted to be an ancient beach deposit. The sandstone contains significant volcanic constituents and ranges in composition from arkosic to tuffaceous. Locally, the host sandstone contains hard, dense opal cement. Bentonitic shales underlie and overlie the mineralized sandstone unit.

Geometrically, the Hobson uranium-bearing zone resembles an elongated roll front type deposit. Chemical uranium values compared with gamma ray equivalent uranium values indicate that the uranium has been dissolved from the updip side and precipitated on the downdip side of the roll front.

Details of the uranium mineralization were studied using thin section microscopy, autoradiography, x-ray diffraction (XRD), scanning electron microscopy (SEM) and energy dispersive spectroscopy (EDS) techniques. The uranium mineralization occurs in the pores of the sandstone, as grain coatings and as replacement of sand-size and smaller grains. It is commonly associated with pyrite and other opaque minerals. Other favored hosts for uranium (other than pore openings) appear to be clay aggregates, altered rock and glass fragments, and carbonaceous plant debris. Aggregates of coffinite were identified by XRD and SEM/EDS analyses. Some uranium mineralization also appears to be dispersed and is postulated to be adsorbed on the clay and carbonaceous matter.

[1]Texaco divested all interest in the Hobson properties and made assigment to Everest Mineral Corporation effective August 16, 1984.

INTRODUCTION

Texaco was active in uranium exploration and mining in the 1950's. In the 1970's, in situ solution mining techniques, for the recovery of more deeply buried low-grade ore deposits, were evaluated as a viable alternative to costly and more conventional methods of open pit and underground mining.

Geologic studies of samples from the uranium exploration holes drilled in the late seventies on the Hobson prospect in Karnes County, Texas were made to determine the amenability of the uranium deposit towards solution mining. The geologic aspects of the studies are presented in this paper. Radiometric relationships and lithologic correlations are developed. Results of detailed mineralogic and petrographic studies of selected core samples are given. Techniques such as autoradiography, x-ray diffraction, scanning electron microscopy and energy dispersive (elemental x-ray) spectroscopy were used to define the uranium mineralization and micro habitat and are discussed in the text.

Location

Hobson field, indicated by the arrow in Figure 1, is located in Karnes County along the south Texas uranium trend. The black dots indicate uranium mineralization. The dots surrounded by circles indicate areas of in situ uranium recovery projects as of 1978.

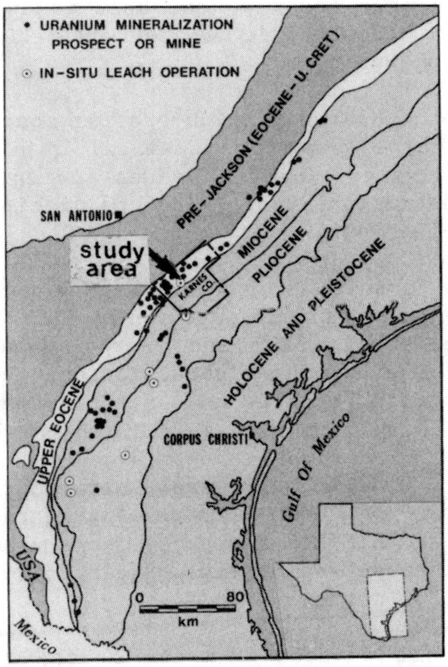

Fig. 1 SOUTH TEXAS URANIUM AREA

Geologic Setting

The uranium project occurs at a depth of 106 to 122 m (350 to 400 ft) in the Deweesville Sandstone, one of several sands that are members of the Whitsett Formation of the Eocene Jackson Group. The mineralized sandstone occurs in sediments that dip gently to the east-southeast. Most of the uranium deposits in the Hobson and adjacent areas (Figure 2) occur as roll front deposits and tend to be aligned parallel to the strike (Dickinson, 1978). Exceptions occur in channel sands, such as the fluvial Kellner channel, which locally cuts out the sheet-like beach sands.

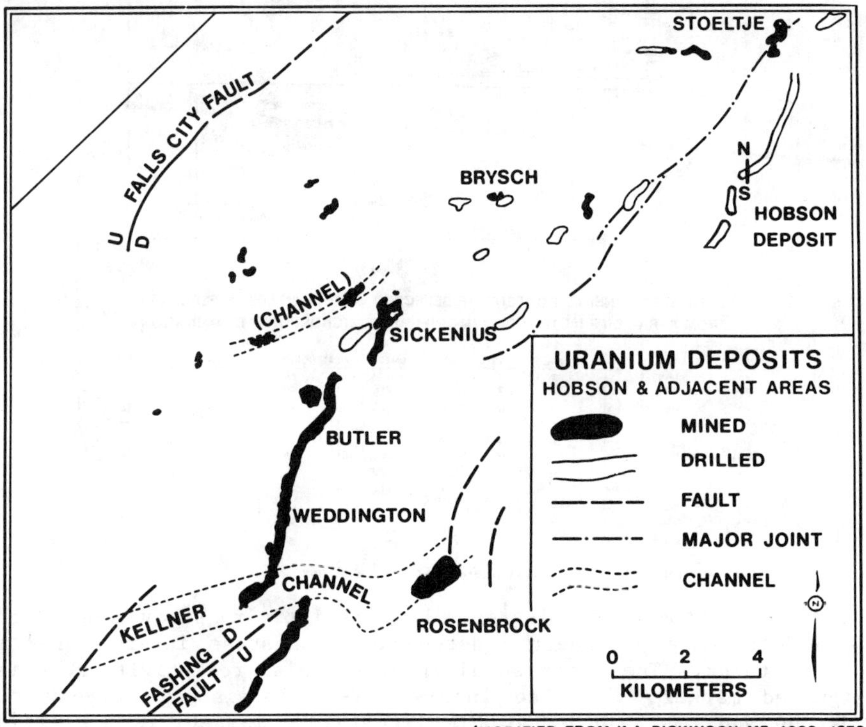

Fig. 2 (MODIFIED FROM K.A. DICKINSON MF-1029, 1978 AND STATE OF TEXAS PUBLIC FILE DATA)

STRATIGRAPHY AND RADIOACTIVITY OF MINERALIZED ZONE

Natural gamma and resistivity logs for a typical well (Figure 3) illustrate several lithologic related features used to construct cross sections and for isopach data. S.P. was run but was of little value due to the invert mud system used. A prominent gamma ray peak "T" coincides with the top of a ligniferous unit "L" showing a high resistivity. Marker "T" has a gamma ray peak of 90-150 CPM on the wells studied. Logs were hung on marker "T" to construct the cross sections (Figures 4 and 5).

Fig. 3 GAMMA RAY AND RESISTIVITY LOGS OF MINERALIZED CORE HOLE
LITHOLOGY, MARKER BEDS AND SAMPLE LOCATIONS INDICATED

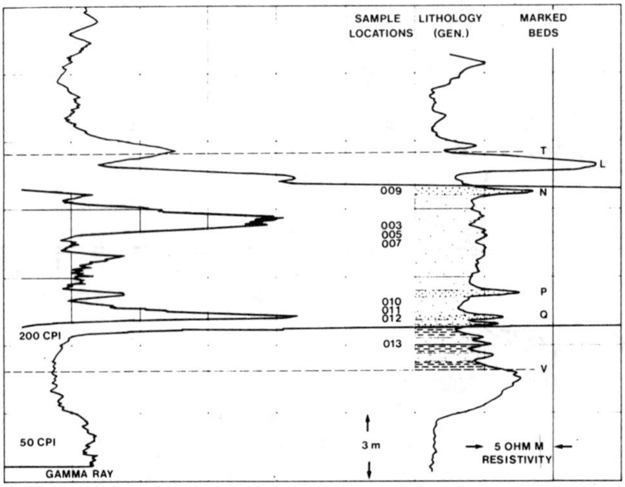

ESTIMATED MINERAL CONTENT OF SELECTED CORE SAMPLES IN PERCENT FROM X-RAY DIFFRACTION DATA MODIFIED BY MICROSCOPIC EXAMINATION

SAMPLE NO.	LITH.*	GLASS	OPAL	CRIST.	CLINOP	SMECT.	MICA	QUARTZ	K SPAR	PLAG.	PYRITE
009	SS-OP	–	8	9	3	1	1	54	9	14	1
003	SD	42	–	2	–	21	1	20	6	7	1
005	MS	11	–	10	–	36	1	26	7	9	TR
007	SD	66	–	2	–	22	1	5	2	2	–
010	SD	–	–	–	3	20	1	48	12	16	–
011	MS-OP-SDY	–	12	12	6	28	1	25	5	8	3
012	SS-OP. SD	–	10	7	13	23	1	25	6	9	6
013	MS	4	8	9	5	24	1	30	7	8	4
101	SD	–	–	–	2	11	1	60	11	15	–
104	SD	–	–	–	2	14	1	56	11	16	TR
106	SS-OP	–	7	13	3	–	–	56	10	14	1

*SS = SANDSTONE; OP = OPALIZED; SD = SAND; SDY = SANDY; MS = MUDSTONE

Near the bottom of the log is a high resistivity sandy zone which is overlain by a prominent mudstone and sandstone interval 1-2 m (3-6 ft) thick. The upper shoulder of the high resistivity zone is designated marker "V". The interval V-T, consisting largely of sandstone and overlain by a thin ligniferous member, is a mapable depositional unit. This unit varies from 9.5 to 11.5 m (29 to 35 ft) thick and contains the porous mineralized sandstone unit. The top of the porous sandstone is generally bounded by a high resistivity bed, marker "N". This marker bed consists of dense, hard opalized sandstone and varies from 0.1-0.6 m (0.3-2 ft) in thickness.

The base of the porous sandstone interval is generally bounded by another dense opalized sandstone, marker "Q". Where these opalized sandstone markers, N-Q, are present, the thickness of the porous and potentially mineralized sandstone unit may be readily defined. In places, dense opalized sandstone is developed within the porous sandstone unit, for example event "P".

Details of the roll front, as defined by gamma ray intensity, are shown on oblique-to-the-strike cross section of four wells, spaced about 32.8 m (100 ft) apart (Figure 4). The approximate location of this N-S cross section is indicated on Figure 2. Figure 4 also includes an interpretative gamma ray cross section. From these profiles, a reasonably good roll front mineralization shape can be inferred. The resulting gamma ray intensity cross section appears as a series of overlapping envelopes folded near the middle and bent back on itself, much like a series of steep sided V's placed inside one another and lying on their sides. The mineralized cross section also resembles a notched wedge with a flat base and slightly convex top.

Roll fronts are generally described as "C" shaped in cross section. That is, in part, caused by vertical exaggeration in illustrations. It should be noted that there is a 2X vertical exaggeration on the cross section in Figure 4. However, this section is cut at about a 60° angle to the dip. Consequently, the vertical and horizontal dimensions of the cross section tend to approximate those of a true dip section.

Fig. 4 HOBSON URANIUM DEPOSIT
GAMMA RAY-RESISTIVITY CROSS SECTION

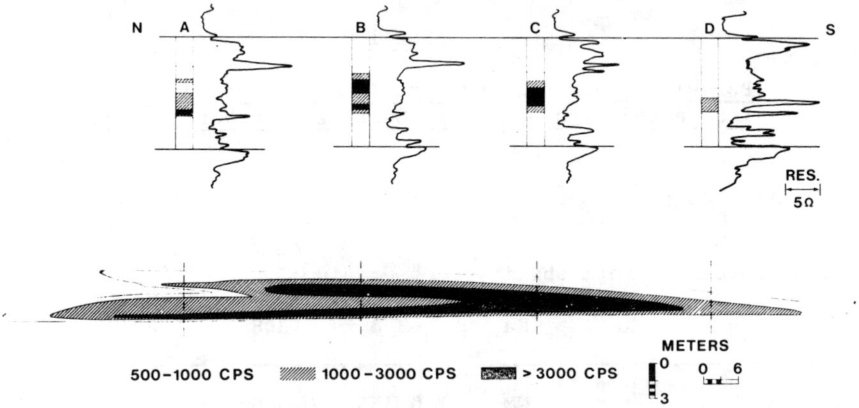

INTERPRETIVE GAMMA RAY CROSS SECTION

RELATIONSHIP OF CHEMICAL TO EQUIVALENT URANIUM

Gamma logs are widely used in the industry to indicate uranium content. However, only a few percent of the gamma photons counted are due directly to uranium. The remainder are due to the daughter products of uranium such as lead 214, protactinium 234 and bismuth 214. In the oxidized state, uranium is readily removed by solution whereas the daughter products are not. Consequently, the daughter products tend to lag behind the uranium. The radioactivity of the daughter products will gradually disappear if they are not replenished by fresh breakdown of uranium.

Roll front deposits are dynamic systems formed in response to oxidation-solution and reduction-precipitation. Consequently, it is possible to define the development of the ore body by a comparison of the gamma ray or equivalent uranium percentage (eU_3O_8), from gamma logs or closed can analysis, with that of the actual or chemical uranium percentage. The ratio of the actual to equivalent uranium percentage is the disequilibrium ratio:

$$D = \frac{U_3O_8}{eU_3O_8}$$

A ratio less than one indicates a deficiency of uranium (leaching) and greater than one indicates an excess of uranium (recent accumulation with insufficient time for the daughter products to have reached equilibrium). Table I contains uranium disequilibrium ratios for selected high-grade ore samples, physical properties and uranium and carbon content. All but one of the samples contain significantly more chemical than equivalent radioactive U_3O_8, indicating a geologically young uranium deposit.

TABLE I

Porosity, Permeability, Uranium and Organic Carbon of Selected High-Grade Ore Samples from Various Wells

Sample No.	Porosity %	Permeability Horiz Ka	Permeability Vert Ka	Uranium % U_3O_8	Uranium % eU_3O_8	Disequilibrium Ratio $D = \frac{U_3O_8}{eU_3O_8}$	Organic Carbon %
1	NM	NM	NM	0.563	0.340	1.66	0.34
2	31.5	2344	1405	0.467	0.281	1.66	0.06
3	32.9	3104	2115	1.183	0.749	1.58	0.05
4	24.5	1347	347	0.469	0.323	1.45	0.25
5	32.1	849	109	0.249	0.177	1.41	0.13
6	35.4	1875	644	0.166	0.108	1.54	0.10
7	27.9	752	165	0.202	0.123	1.64	0.89
8	35.2	3959	2252	0.588	0.724	0.81	0.51

NM = Not Measured

Figure 5 contains the chemical and equivalent uranium content for the four wells described in Figure 4. This cross section, hung on marker bed "T", is on the same scale as the corresponding gamma ray-lithology cross section in Figure 4. Note the general wedge shape of the uranium mineralization pointing in the downdip direction (towards well D). The thick zones of mineralization in the updip wells A and B exhibit a deficiency of uranium content in the lower middle zone of mineralization (corresponding to the core) and an excess of uranium in the upper and lower limbs. Note also that the gamma ray (daughter product) radioactivity in the core zone is less in the updip well A. At well C, the mineralized zone is located in the lower portion of the sandstone and is thinner, of higher grade and has an excess of uranium. The most downdip well, D, has weak mineralization but has an excess of uranium in the zone correlatable with well C.

The geometric relationships of the mineralized zone combined with the positions and intensities of the uranium excess and deficient zones indicate that the mineralization moves both downdip to form the core or leading edge, and has an upward and downward component within the sand to form the wings. Oxidation and solution are greatest in the lithologic zone corresponding to, and updip of, the mineralized core. Still further updip, the limbs are oxidized and the uranium solubilized.

URANIUM CONTENT OF CORES (CHEMICAL AND EQUIVALENT)
RESISTIVITY CROSS SECTION

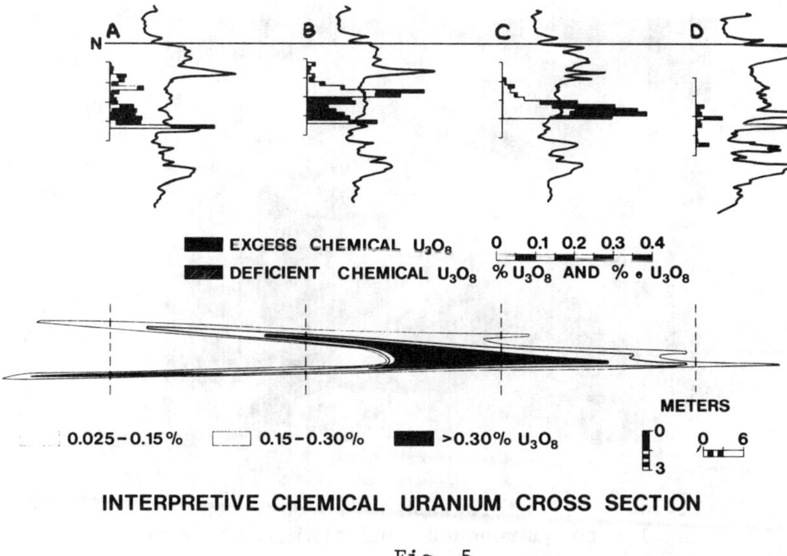

INTERPRETIVE CHEMICAL URANIUM CROSS SECTION

Fig. 5

The example given is a textbook example of a uranium roll front deposit. Many are not so simple. The relationships developed above may be used to explore for ore bodies and may be applied to uranium recovery programs such as those involving in situ solution leaching. For example, the ore body, as defined by chemical core analyses, may be ten meters downdip from the ore body as defined by gamma ray logging. The chemical environment of the updip, partly oxidized mineralized zone would be markedly different from the downdip, reduced ore zone. Also, optimization of uranium recovery from the depleted updip, partly oxidized zone may require different recovery techniques than for the reduced zone.

MINERALOGY AND PETROGRAPHY

The estimated mineral content of selected core samples is shown in Figure 3. Mineral identifications were derived from semi-quantitative x-ray diffraction analysis techniques and modified by results of petrographic examination of thin sections and binocular microscope examination of cores. The latter is required due to the significant amount of volcanic glass in these samples. Glass, as well as opal, is amorphous and not amenable to x-ray diffraction analysis.

SEM and petrographic examinations reveal the samples are fine- to very fine-grained, subangular to subrounded, well sorted, weakly cemented, porous sandstones (Figure 6). The sandstones are characterized by a high content of first cycle volcanoclastic components including glass shards, volcanic rock fragments and feldspars.

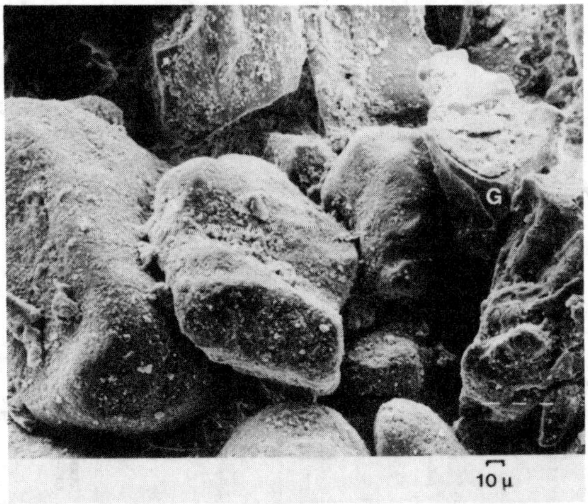

Fig. 6. SEM micrograph of porous, subangular to subrounded sand grains. Note angular glass shard (G).

Petrographic examination reveals that the volcanic glass shards are generally clear to brownish, and are largely unaltered (Figure 7). Glass content ranges from not detected to 66%. It is present in most samples. The samples contain moderate percentages of both plagioclase (2-16%) and K-feldspar (2-12%). Most of these grains are angular to subangular and appear relatively unaltered, typical of young volcanic detritus. Some embayed quarts grains were observed, attesting to their volcanic origin. However, most of the quartz grains are subangular and appear to be derived from reworked sedimentary rocks. The small amount of mica, identified by XRD and petrography, appears to be largely biotite, which forms pseudohexagonal flakes.

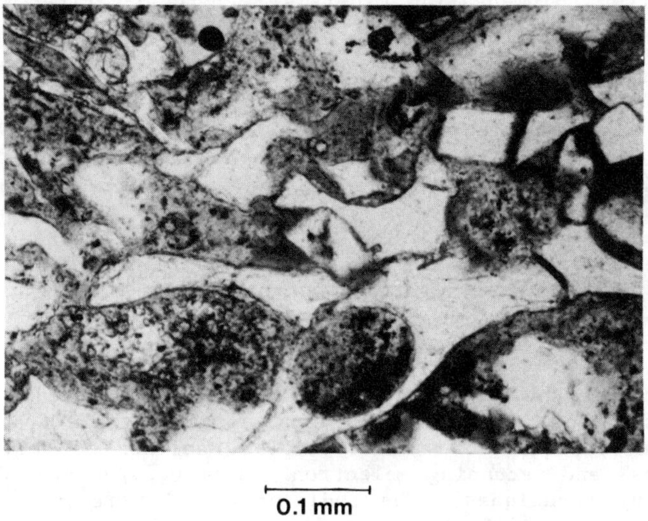

Fig. 7. Thin section photomicrograph of porous, tuffaceous sandstone.

The sandstones are largely uncemented or are weakly cemented with montmorillonite, opal and/or the zeolite, clinoptilolite. Some of the sandstone has been tightly cemented with opal and also with finely crystalline cristobalite and tridymite (opal-CT), resulting in a hard, dense siliceous sandstone. Clinoptilolite occurs as minute laths, sparse pore fillings and alteration products.

Montmorillonite comprises about 20% (with less in many cores) of the mineral content and occurs as clay pellets, vesicle fillings, wholly or partially enclosed in the volcanic glass shards and as alteration products of the glass shards, feldspars and rock fragments.

Montmorillonitic alteration ranges from thin, discontinuous surface coatings to complete argillic alteration of the clastic grains. Only minor quantities of montmorillonite occur as pore fillings. The montmorillonite contains no x-ray identifiable interlayered clays and readily expands from 15.4 A at 50% relative humidity to 17 A upon glycolation. This indicates that two molecular water layers are present and suggests that the montmorillonite is calcium-saturated.

All samples from the downdip, reduced ore zones contain a small amount (up to 6%) of pyrite, which forms a fine-grained replacement of woody fragments and iron-bearing minerals. Opaque grains containing fine disseminated pyrite appear brassy yellow in reflected light. Other opaque grains (in some samples) appear white in reflected light and are believed to contain leucoxene, an alteration product of titanium-bearing minerals such as ilmenite, rutile and sphene. Some unaltered parts of clay pellets, and argillic alteration of some of the non-opaque grains also appear white in reflected light. Samples retrieved from wells in the updip, oxidized zones were largely devoid of pyrite.

For the most part, carbonate minerals are notably absent. XRD identified minor amounts of dolomite and/or calcite in a few samples that slightly effervesced when treated with a 10% solution of hydrochloric acid. Minor amounts of iron-rich carbonates, siderite and/or ankerite, were identified by petrographic examination and appear to be associated with the montmorillonite-rich layers.

URANIUM MINERALIZATION

Details of the uranium mineralization and distribution were studied using both radioluxography and alpha particle tract autoradiography techniques, x-ray diffraction analyses of heavy mineral separations and scanning electron microscopy/energy dispersive spectroscopy techniques. The radioluxographs were made with zinc sulfide phosphor 1101D, Polaroid film, type 52 and 3 days exposure time. The resulting radioluxographs produced a diffuse pattern of light areas corresponding to radioactive zones.

A more definitive method of tracing radioactivity involves uncovered thin sections (polished rock slice 30 microns thick on a glass slide) placed against alpha-sensitive nuclear track plates (emulsion thickness 50 microns on a glass slide) for longer exposure time (20 plus days). To facilitate microscope study, the matched pairs were then oriented with the rock slice on the microscope stage and the emulsion towards the objective. This allows one to focus on the alpha particle track plates or the thin section without undue interference from the other. However, the same side of the rock slice which was originally in contact with the nuclear emulsion is not the same side in focus.

The most concentrated alpha particle tracks, indicative of uranium mineralization, are associated with some grains containing pyrite (and/or marcasite) and black opaque minerals. These strongly radioactive grains appear to have been originally detrital heavy minerals, iron-rich silicates, clay vesicle fillings, brownish clay aggregates and carbonaceous plant debris. Details of a group of radioactive opaque grains are shown in Figures 8a,b and c. The dark minerals appear to be largely brass yellow in reflected light, indicative of pyrite and/or marcasite. The sulfidization appears to be a fine-grained replacement of preexisting grains. The concentrations of alpha-particle tracks indicate that there are significant concentrations of uranium-bearing material intimately intermixed with the iron sulfides. Diffuse radioactivity is observed in other samples and appears associated with clay mineral aggregates, partly altered rock fragments, altered glass shards and plant debris. Although intense concentration of alpha particle tracks are derived from opaques, pyritized clay pellets and carbonaceous plant debris, not all of the pyrite concentrations and plant debris exhibit anomalous radioactivity.

Total percentage organic carbon (predominantly woody plant debris) was determined for the previously selected samples (Table I). The organic carbon content ranges from 0.05 to 0.93% and averages 0.30%. A comparison of the uranium assay (%U_3O_8) with the organic carbon content reveals no correlation between them. It would appear that there is sufficient organic carbon and/or other reductants to reduce the uranyl ion even in the carbon-poor sands.

The association of pyrite with concentrations of alpha particle tracks suggests that uranium-bearing constituents are concentrated in the heavy mineral fraction. Heavy mineral separations (bromoform method, Sp.Gr. approximately 2.8) were made of samples from selected core holes. Magnetic and non-magnetic fractions were obtained by means of the Frantz Isodynamic Separator set at 1 amp. The resulting heavy mineral composition was determined by x-ray diffraction analysis. Mica, predominantly brown volcanic biotite flakes, is abundant in the magnetic separates but does not occur in the non-magnetic fractions. Other minerals such as titanium-bearing ilmenite and rutile tend to be concentrated in, but not restricted to, the magnetic separates. Pyrite, less abundant marcasite, quartz, zircon and tourmaline are more concentrated in the non-magnetic fractions. Coffinite, the hydrous uranyl silicate, was identified in the non-magnetic fraction. Uraninite was looked for but was not positively identified. The overlap of the strong diffraction peaks of uraninite makes the identification of small amounts of uraninite in the presence of pyrite difficult.

Figure 9 is an SEM micrograph that illustrates the common occurrence of montmorillonite as a thin, continuous substrate on a partially altered glass shard. The small rounded aggregates are opal. The laths are clinoptilolite. The uranium mineralization is

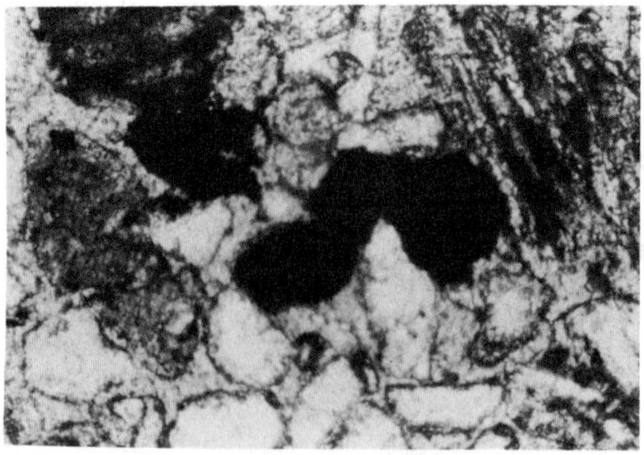

Fig. 8a. Radioactive grains (black) are replacement pyrite intermixed with a uranium mineral.

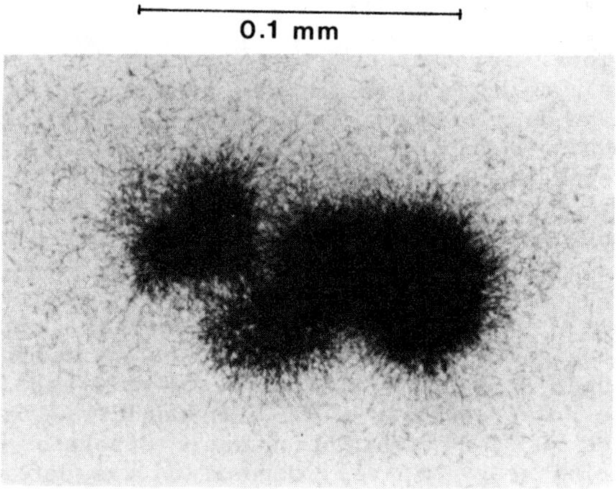

Fig. 8b. Audioradiograph

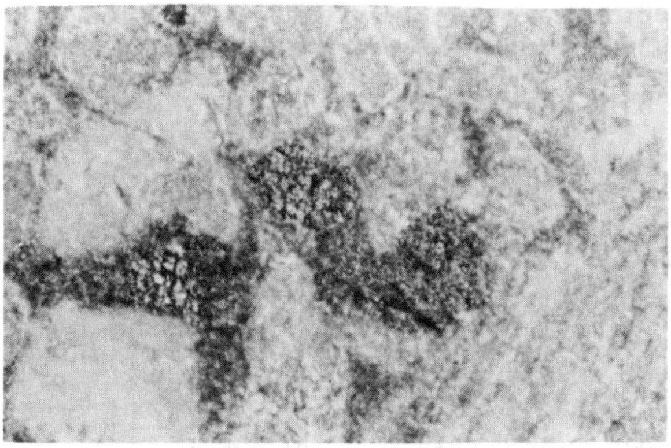

Fig. 8c. Inverted thin section in reflected and transmitted light.

suspected to form minute aggregates of poorly developed crystals (arrows) affixed to the montmorillonite. Figure 10 is the EDS elemental analysis of the uranium-bearing clusters and proximal substrate. The spatial limitations of the electron beam prevent localized analysis of such minute (less than 1 micron) subjects. Interpretation of the EDS spectrum, based on the intensity of the elements, silicon and uranium, is the mineral coffinite.

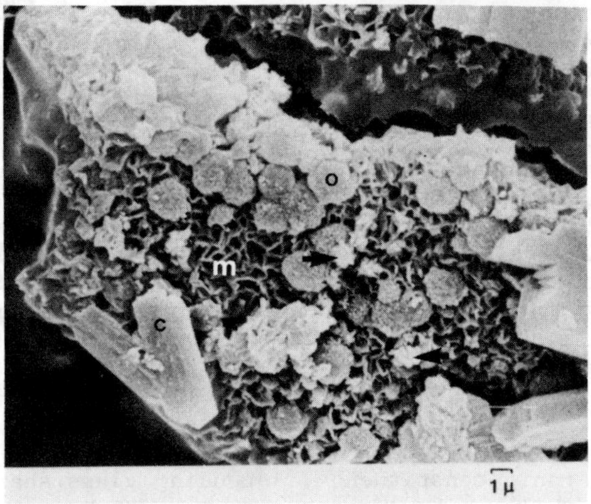

Fig. 9. SEM micrograph of partially altered glass shard with montmorillonite (m) substrate, opal cement (o), clinoptilolite laths (c) and coffinite (arrows).

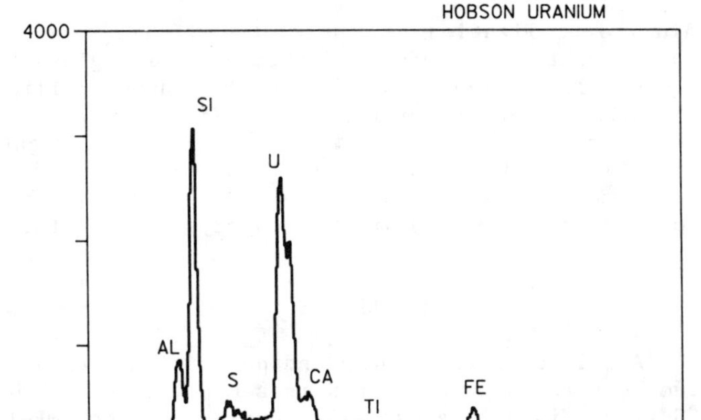

Fig. 10. EDS Spectrum. Intense uranium and silicon peaks suggest the mineral coffinite. Other elements and some of the silicon may be related to proximal minerals.

The occurrence of pyrite (FeS_2), ilmenite ($FeTiO_2$) and rutile (TiO_2) with uranium has been described by Reynolds, et al., (1977) in an unidentified south Texas roll-type ore body. They noted that uranium is found in association with a TiO_2 phase which originates from the expulsion of Ti during the replacement of Fe-Ti oxide minerals by pyrite. TiO_2 is a known scavenger of uranium and has been proposed as a sieve material to recover uranium from sea water. Reynolds, et al., also noted the occurrence of uranium in the light mineral fraction with silicon, calcium, potassium and aluminum. This would appear to coincide with the chemical constituents found in clay minerals, rock fragments and feldspar.

CONCLUSIONS

The Hobson uranium prospect occurs as a typical roll front deposit in a sandstone and mudstone unit. The roll front in cross section resembles an inclined wedge with a notch at the back (a core or front with trailing wings). The strongly mineralized core, located on the downdip side, has excess chemical uranium over gamma ray equivalent uranium. The oxidized updip core and oxidized updip extremities of the limbs have a deficiency of chemical uranium versus gamma equivalent uranium.

The Hobson sandstone cores are generally fine-grained, well-sorted and porous, but some are cemented and tight. A large percentage of volcanic constituents, including glass shards, feldspars and montmorillonite, are often present. Sandstones are largely uncemented or weakly cemented with opal, montmorillonite and/or clinoptilolite. Some of the sandstones have been opalized to hard, dense units and appear to be barriers to vertical flow.

The uranium mineralization is associated with pyrite and other opaque minerals, clay aggregates, altered rock and glass fragments and carbonaceous plant debris. Coffinite, a uranous silicate, was identified by XRD and SEM/EDS techniques. SEM/EDS analyses suggest the coffinite forms minute (less than 1 micron) aggregates that affix to thin montmorillonite substrates. Some of the uranium mineralization is very weak and dispersed. This may be due to adsorption of uranium, probably in the non-crystalline state, on the clay and carbonaceous materials.

REFERENCES

Dickinson, K. A., 1978, Stratigraphy and depositional environments of uranium host rocks in western Karnes County, Texas," U.S. Geological Survey, Miscellaneous Field Studies Map, No. MF-1029.

Reynolds, R. L., Goldhaber, M. B., and Grouch, R. I., 1977, "Uranium associated with iron-titanium oxide minerals and their alteration products in a south Texas roll-type deposit," Short Papers of the U. S. Geological Survey, Uranium-Thorium Symposium, 1977, Geological Survey Circular No. 753.

CLAY MINERAL INVESTIGATIONS

Chapter 8

THE SIGNIFICANCE OF CLAY MINERALOGY IN THE AMENABILITY OF SANDSTONE VANADIUM ORES

by D. M. Hausen

Newmont Exploration Limited
44 Briar Ridge Road
Danbury, CT 06810

ABSTRACT

The amenability of a given vanadium ore to any given treatment depends largely on the mineralogic combinations of vanadium in the ore. Quantitative data on vanadium mineralogy provide not only an objective basis for selection of proper treatment, but also may eliminate many time-consuming amenability studies whose results are empirical, at best. Vanadium is distributed through the interstices of mineralized sandstone ores in the form of oxidized vanadates, reduced oxides, and silicates. Oxidized vanadates, e.g., carnotite, tyuyamunite, etc., are most readily beneficiated by simple acid leaching. Reduced oxides are more slowly solubilized, but relatively amenable to most acid treatments. The layered silicates, roscoelite, mixed layered mica-montmorillonite, and chlorite, are least amenable to most acid extractions, and constitute an assemblage of vanadium-bearing clays that are common to many sedimentary deposits. Vanadium is combined principally as layered silicates in the clay fractions of many sandstone ores. Mineralogic studies in this paper are largely confined to clay minerals, and to their amenability characteristics in low pH liquors.

INTRODUCTION

A basic knowledge of vanadium mineralogy, particularly vanadium clays, is necessary in the evaluation of amenability data from different type ores.

Vanadium is distributed interstitially through the matrices of mineralized sandstones in the form of oxidized vanadates, reduced oxides, and silicate phases. Oxidized vanadates may become quite complex in composition, but usually include such common phases as carnotite, tyuyamunite, and zippeite, which are readily solubilized and recovered by direct acid leaching. On the other hand, reduced oxides, including montroseite, paramontrosite, corvusite, etc., are more slowly dissolved, but relatively amenable to most oxidizing acid treatments.

The layered silicates, including roscoelite, mixed layered mica-montmorillonites, and chlorite, are least amenable to acid extractions, and constitute an assemblage of vanadium-bearing clays that are common to most Salt Wash and Entrada ores on the Colorado Plateau. Detailed mineralogic studies have indicated that vanadium is combined principally as layered silicates in the clay fractions of most vanadium ores. Reduced oxides and vanadates, although locally predominant in select high grade samples, are usually secondary in abundance (and economic significance) to vanadiferous clays which comprise the major mineralogic source of vanadium in most bulk shipments of ore from the Uravan Mineral belt.

For the above reasons, metallurgical treatments must be designed to release and extract the vanadium most efficiently from the lattices of clay silicates. For example, salt roasting is an effective pyrometallurgical treatment for vanadium clays because the silicate lattice breaks down near 850°C to form a high temperature sodium feldspar near albite (Hausen, 1983) and the excess sodium from NaCl combines with the released vanadium to form soluble $NaVO_3$.

However, the extraction of vanadium from such clays by means of acid leaching is not so simple. Special emphasis is given in this paper to the dissolution of vanadium-bearing clays by low pH sulfate liquors.

Acknowledgments are extended to the Union Carbide Corporation and to Newmont Mining Corporation for their permission to publish the findings of this preliminary study.

MINERALOGY AND BACKGROUND OF VANADIUM-BEARING CLAYS

A variety of clay minerals occur with uranium and vanadium mineralization in Salt Wash and Entrada sandstones on the Colorado Plateau. Subtle differences in clay mineralogy have been noted in ore shipments from different localities, and are reflected in variable physical and chemical characteristics to plant processing. In some instances, mineralogic differences are known to account for variations in amenability to acid leaching. Relationships between clay mineralogy and vanadium recovery in pyro- and/or hydrometallurgical processing is rather poorly understood, and requires integrated mineralogic-metallurgical study before amenability results can be evaluated on the basis of mineralogical data. Mineralogical data on clays were accumulated from various Union Carbide properties over a three-year period between 1959 and 1961; preliminary results from this study were sufficiently encouraging to warrant further study.

Several different species of clay minerals (Table I) have been identified from the x-ray diffraction analysis of mine and mill samples, and correlated with location of company properties in Colorado and Wyoming (Fig. 1).

The following two significant differences were noted between clays from the Mesozoic ores of the Colorado Plateau and the Tertiary ores of northern Colorado and Wyoming: (1) the abundance of mixed layered mica-montmorillonites in the former, and (2) the lack of vanadium in the latter. Most of the vanadium in Salt Wash ores is contained in mixed layered mica-montmorillonites. Some vanadium is also contained in chlorites and roscoelites. Clays from different localities may vary in mineralogy sufficiently to require different metallurgical treatments for optimum vanadium recovery.

Mixed layered mica-montmorillonites are the most widespread and abundant group of clay minerals in the Uravan Mineral Belt. The interstratification of montmorillonite within clay mica layers affects the solubility of vanadium from this important vanadium mineral, and can be used as an index to predict the amenability of certain types of ore feed.

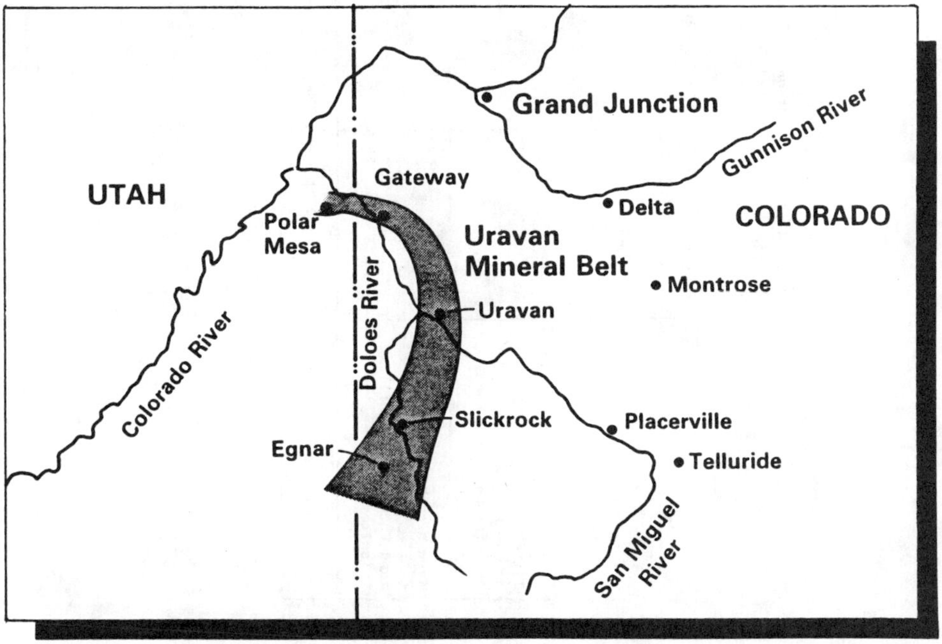

Fig. 1. Locality Map of Uranium Ores from Slick Rock and Placerville areas, south of Uravan, Colorado

TABLE I. Mineralogic Distribution of Clays in Union Carbide Properties in Colorado and Wyoming

Clay Mineral	White River from Riverton, Wyoming	Brown's Park from Maybell, Colorado	Morrison from		Entrada from	
			Uravan	Slick Rock	Rifle	Placerville
Kaolinite	X		x-X	x		
Montmorillonite	X	X				
Chlorite			x	X	x-X	
Illite		x				x
Vanadiferous Illite and Roscoelite			x		x-X	X
Mixed Mica-Montmorillonites			X	X	x	
Mixed Chlorite-Montmorillonites			x	x		

x Minor X Major x-X Minor to Major

MIXED LAYERED CLAYS

General Background

Mixed layered clays are perhaps the most widespread and, until recently, the least understood variety of clays in sedimentary rocks. Studies by C. Weaver (Weaver, 1955, 1956) have indicated that mixed layered clays occur in 70 percent of sedimentary rocks. The most abundant varieties are randomly interstratified clays of illite-montmorillonite and illite-chlorite-montmorillonite. Mixed layered chlorite-montmorillonite and chlorite-vermiculite are also common, but relatively less abundant.

Mixed layered illite-montmorillonite is the predominant variety of interstratified clay in the Salt Wash ores of the Uravan Mineral Belt. Interstratification of montmorillonite with mica layers has been studied by J. C. Hathaway (Hathaway, 1955, 1959) and M. D. Foster (Hathaway, 1959) of the U.S. Geological Survey. According to J. C. Hathaway, vanadiferous clay micas may resemble roscoelite but usually differ in two ways: (1) interstratification of montmorillonite, and (2) variation in octahedral substitution of vanadium. Increase in vanadium is accompanied by decrease in aluminum in octahedral positions.

Most of the vanadium in ores of low to moderate grade in the Salt Wash are believed to be contained in mixed layered mica-montmorillonites.

Mixed Layering as a Guide to Geologic Interpretation

The identification of mixed layered clays in recent years has vastly increased the criteria used for the characterization and zonation of sediments. According to C. E. Weaver (Weaver, 1956), mixed layered illite-montmorillonites commonly have ratios near 70:30 and 30:70, illite to montmorillonite (I/M). X-ray diffraction studies at Grand Junction, Colorado, have indicated that ratios of 90:10 to 75:25, I/M, are common to clays from Uravan and Slick Rock ores. Some mixed layering of chlorite and montmorillonite has also been determined in Slick Rock ores. The variety of mixed layering in a range of ratios increases the likelihood of using clay minerals as an aid to geologic interpretation, and possibly as a guide to ore.

According to C. E. Weaver (Weaver, 1956), mixed layered clays are derived in most cases from the degradation or aggradation of preexisting clay minerals, and are quite sensitive to environmental changes. These views are consistent with those of M. D. Foster (Foster, 1959), whose studies suggest that unmineralized clays of the Salt Wash become mineralized simply by the replacement of some of the octahedral aluminum by quadrivalent vanadium in mixed layered clays. Although some vanadium replacement may have occurred in interstitial clay minerals, paragenetic studies in Grand Junction indicate that both roscoelite and mixed layered varieties may replace sand grains epigenetically in sandstones (Fig. 2) containing minimum amounts of pre-existing clay. Zonation of mixed layering as a guide to ore in the Salt Wash requires further investigation.

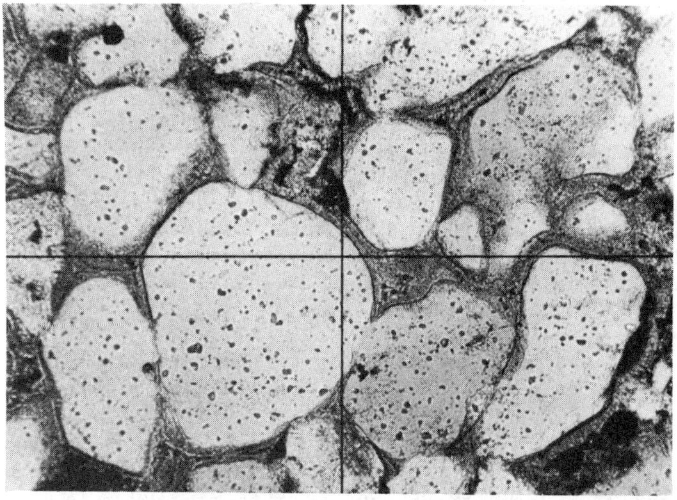

Fig. 2. Coronas of vanadiferous clays, e.g., chlorite and montmorillonites, rimming subrounded quartz grains (Q) in ore sample from mine near Slick Rock, Colorado. Transmitted light, X333.

Mixed Layering as a Guide to Amenability of Ore Samples

Most of the insoluble vanadium in acid leach tails from UCNC ores is octahedrally bound within illitic clays whose silicate structure must be broken down before vanadium can be effectively liberated and solubilized. Clay structures are slowly broken down by prolonged acid treatment, or rapidly destroyed by roasting at elevated temperatures. Differential thermal analyses have been run on Slickrock clays (90:10, I/M) and compared with standard curves for illite, montmorillonite and a 40:10 mixed layered mica-montmorillonite in Figure 3. The thermal curve for 90:10 Slick Rock clay agrees most closely with that of pure illite, and confirms the slight degree of mixed layering of this vanadium clay.

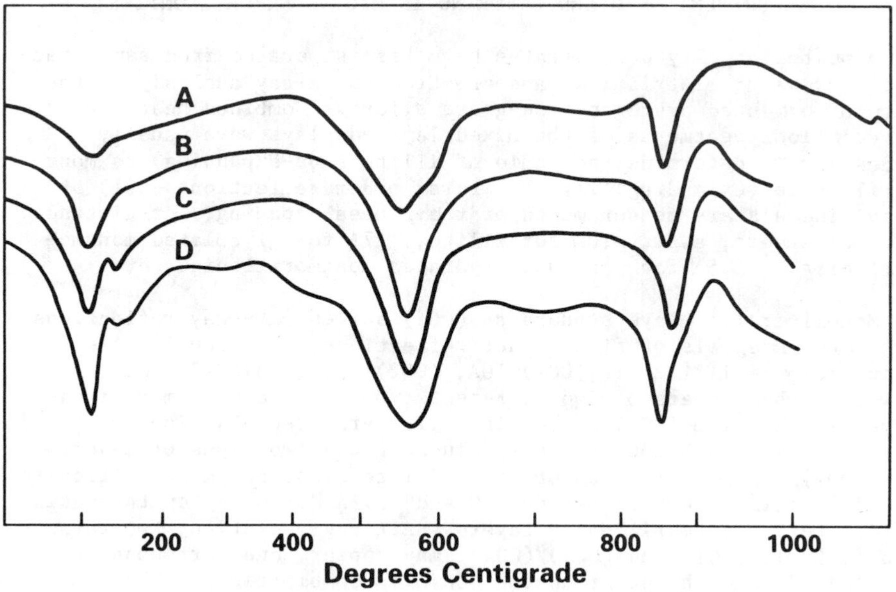

Fig. 3. DTA curve of clay from Slick Rock (A) Compared with standard curves for illite (B), montmorillonite (C), and mixed layered Illite-Montmorillonite (D). [Curves B, C, and D after Grim, Bradley and Brown, (Grim, Bradley, and Brown, 1951)].

Interpretation of the Slick Rock curve indicates that absorbed water is lost between 100°C and 200°C, followed by dehydroxylation at 560°C, and final destruction of the clay lattice at about 890°C. Mixed layering in the range from 90:10 to 40:10, I/M, does not appear to affect the final destruction of the clay near 900°C, and is not believed to be a factor in the amenability of clays to roasting.

On the other hand, the amenability of these clays to acid leaching depends largely on the degree of mixed layering of montmorillonite between illite sheets. The availability of vanadium to leach liquors depends on the number and distribution of expanding layers of montmorillonite between the illite sheets. The degree of mixed layering in mica-montmorillonites can be measured by x-ray diffraction, and is described in the following section.

MEASUREMENT OF MIXED LAYERING IN MICA-MONTMORILLONITES

A number of clay concentrates have been separated from sand fractions by water elutriation, and prepared for x-ray analysis in the form of oriented aggregates on glass slides. Combined heat and glycolation treatments of the mixed layered clays were usually necessary to determine the ratio of illite (non-expanding) to montmorillonite (expanding) layers. First order reflections (001) of clay minerals are measurements of their basal spacing, or distance between sheets, e.g., ∿10A for illite, ∿17A for glycolated montmorillonite, 12.5A for partially hydrated montmorillonite, etc.

Monomineralic clays produce an (00ℓ) series of x-ray reflections that are integrals of first order reflections. The (00ℓ) reflections of pure illite are (001)=10A, (002)=5A, (003)=3.3A, etc. However, when layers of two different types form a randomly mixed-layered clay, the (00) series is no longer integral. The 00 reflections then become average values of the two types of layers. The (001)/(001) reflection of mixed illite (10A) and montmorillonite (12.5A) falls somewhere between 10 and 12.5A depending on the ratio of illite to montmorillonite layers. After glycolation, two reflections, (001)/(001) and (001)/(002), may appear, one corresponding to the 10A/17A and the other to the 10A/8.5A combinations. The 10A/17A reflection gradually approaches 17A, and the 10A/8.5A approaches 8.5A as the percentage of expanded layers of montmorillonite increases.

The theoretical basis for the identification of various mixed layered clays has been developed by Hendricks and Teller (Hendricks and Teller, 1942), Bradley (Bradley, 1945, 1953), MacEwan (MacEwan, 1942), Brown and MacEwan (Brown and MacEwan, 1950) and Brindley (Brindley, 1951), and the reader is referred to the works of these investigators for a more detailed description of mixed layered clays. Two curves are plotted in Figure 4 from data determined by Brown and MacEwan (Brown and MacEwan, 1950) and C. E. Weaver (Weaver, 1956) for the measurement of I/M ratios from 10A/12.5A and 10A/17A reflections. Measurements of illite-montmorillonite ratios in vanadiferous mica-montmorillonites of Salt Wash and Entrada ores are based on these curves.

Three samples of ore from the Burro and Deremo mines near Slick Rock, and old abandoned workings near Placerville, Colorado, were selected for amenability study and were analyzed for clay mineralogy. X-ray diffraction patterns of the three samples are shown in Figure 5. The diffraction peaks for mixed layered clays from the Burro ore are relatively weak, and show the most mixed layering of montmorillonite. Weak intensities were also reported by C. E. Weaver (Weaver, 1956) for clays near 30 percent expanded layers. Diffraction intensities increase appreciably as the percentage of expanded layers decreases below 20 percent. Descriptions and analyses of the clay fractions from each sample are given in Table II.

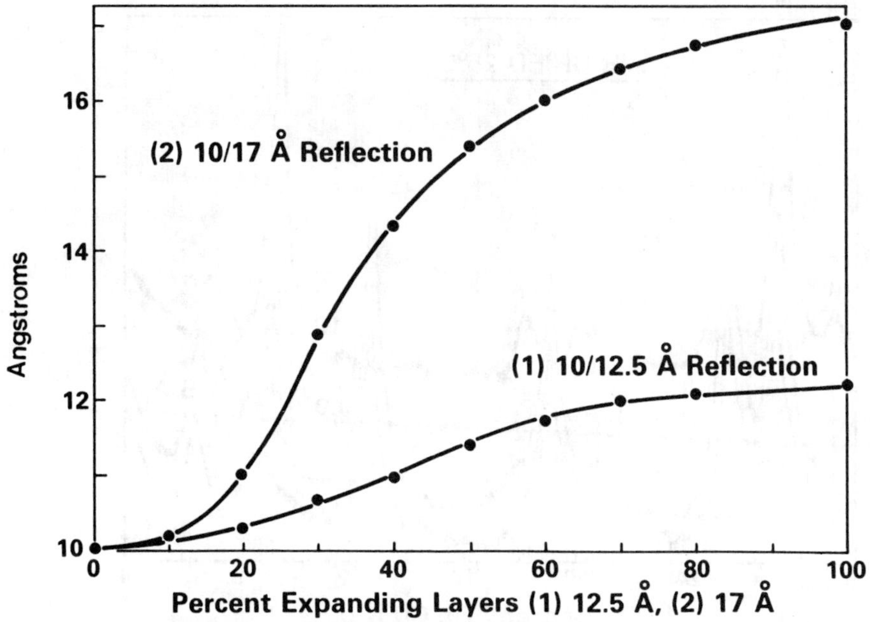

Fig. 4. Curves showing migration of (001/(001) reflections of randomly interstratified 12.5 Å and 17 Å layers between mica layers. (After Weaver, 1956.)

TABLE II. Description and Analyses of Mica-Montmorillonites from Burro, Deremo, and Placerville Samples

Sample No.	Name	Formation	Locality	Percent V_2O_5	Percent Expanded Layers
2633B	Burro Mine Near SlickRock	Morrison (Salt Wash)	Slick Rock District	6.85	20-25 percent
2658	Deremo Mine Near SlickRock	Morrison (Salt Wash)	Slick Rock District	8.77	10-15 Percent
2695	Bear Cr. Mine Near Placerville	Entrada	Placerville District	10.08	0-5 percent*

* Unmeasurable, virtually nil expanding layers.

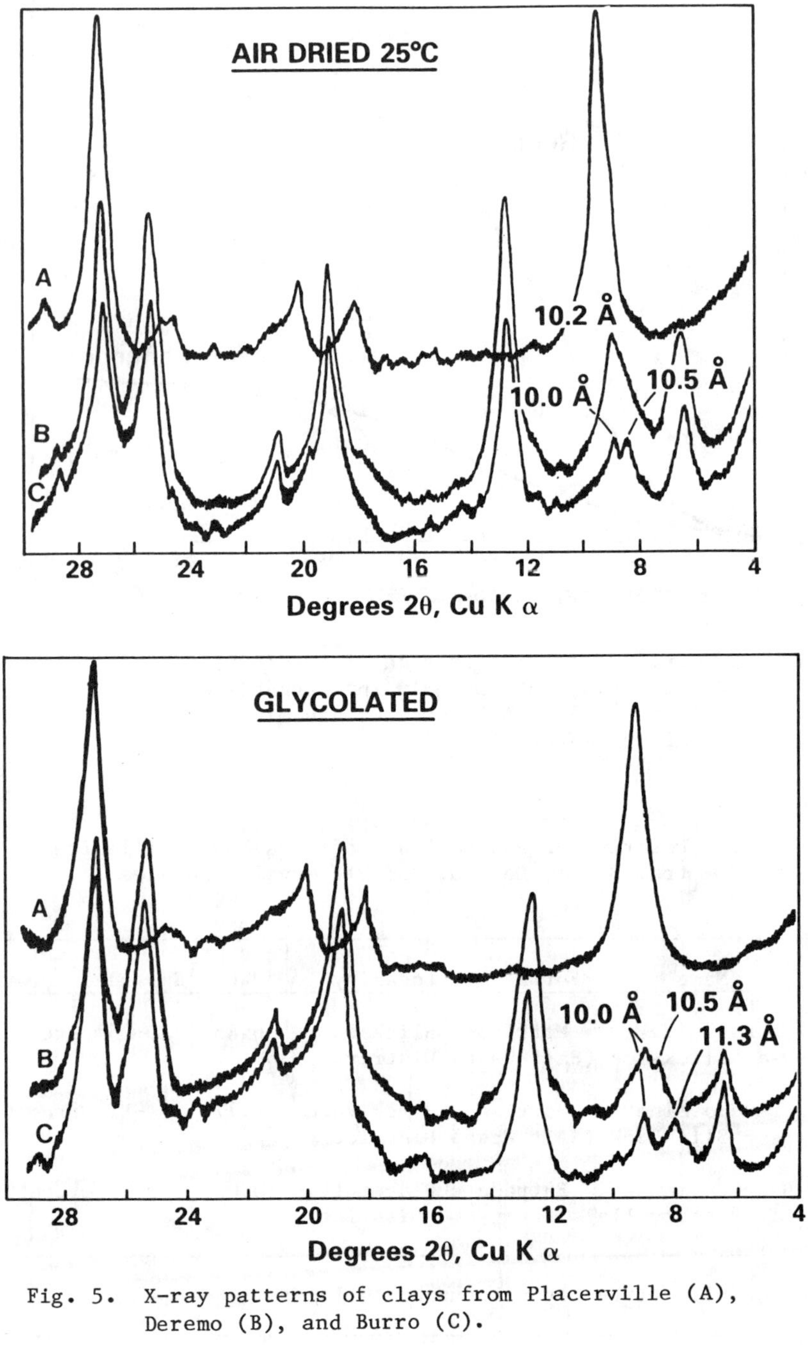

Fig. 5. X-ray patterns of clays from Placerville (A), Deremo (B), and Burro (C).

It is interesting to note that V_2O_5 assays of clays are inversely related to amount of expanded layers in the clay, suggesting that vanadium may be concentrated mostly in the non-expanded illite layers. Vanadium grades are plotted against percent expanded layers in Figure 6. It is recognized that this relationship may hold only for randomly interstratified mica-montmorillonites that are highly mineralized.

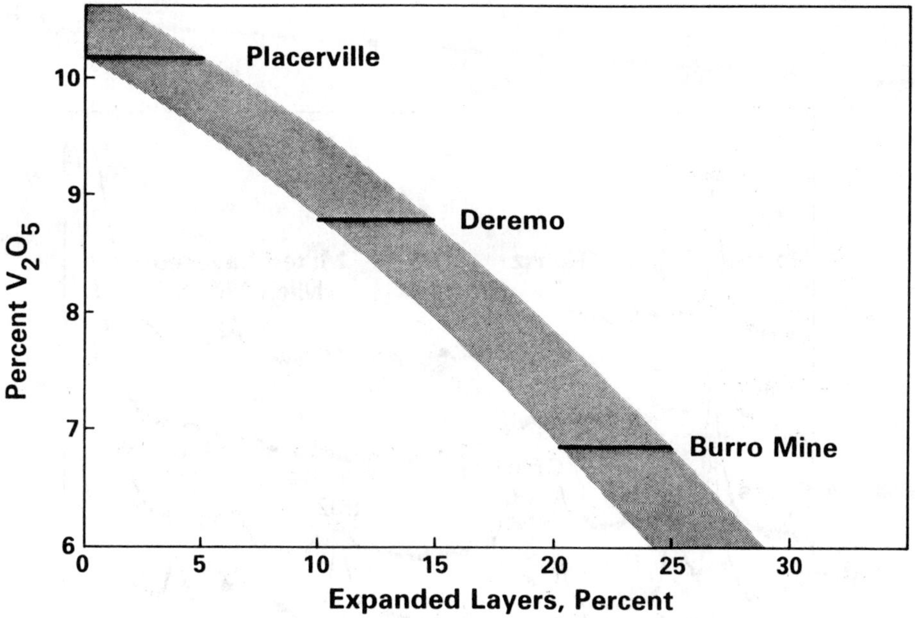

Fig. 6. Inverse relationship of V_2O_5 content to percent expanded layers in head samples of highly mineralized clays.

AMENABILITY OF MICA-MONTMORILLONITES TO ACID LEACHING

Representative portions of the Burro, Deremo and Placerville samples were contacted for periods up to 32 hours at 80°C with sulfuric acid solutions containing 50 percent solids at a pH of 0.5 and e.m.f. of (-) 500. Thief samples were taken at recorded intervals during acid leach tests, and clay fractions were carefully separated by water elutriation from each thief sample for x-ray diffraction and chemical analyses. Sequential decreases in crystallinity and vanadium content of leached clay residues were determined as a function of time of leaching. Gradual breakdown of clay minerals with increasing leach times was noted in x-ray patterns of leached residues (Fig. 7). A removal of vanadium from the clay accompanied the gradual destruction of the clay lattice (Fig. 8).

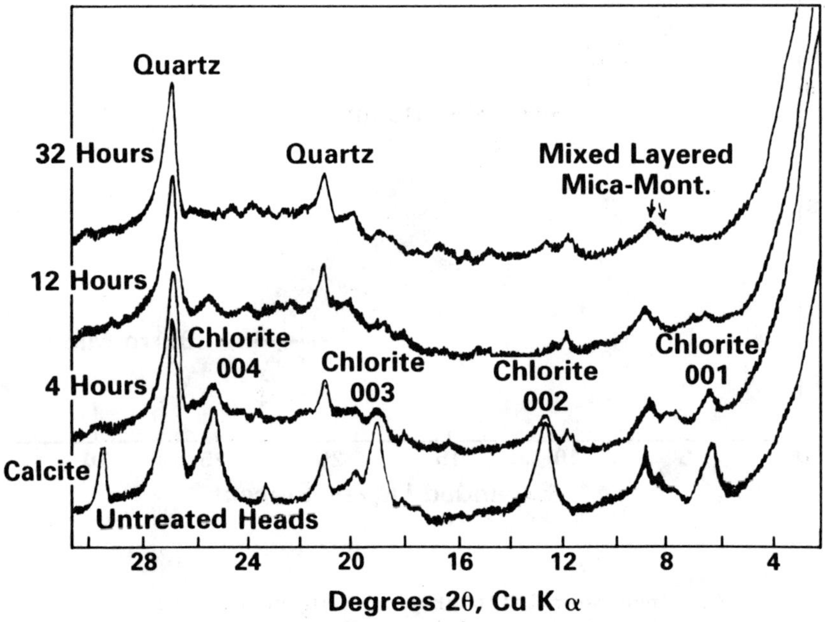

Fig. 7. Destruction of clay structures as a function of acid leaching time (after 4, 12 and 32 hours). Burro ore sample.

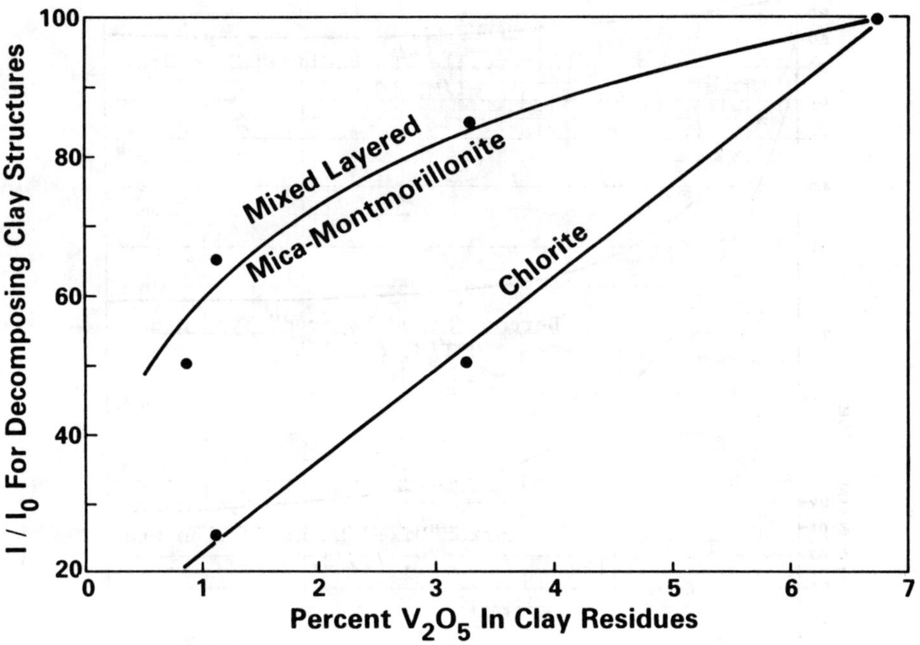

Fig. 8. Correlation of V_2O_5 with decreasing relative I/I_o intensity (arbitrary units) of 001 spacing. for clay minerals in acid-leached residues of Burro ore.

Amenability curves for vanadium (Fig. 9) are distinctly different for each of the samples from the Burro, Deremo, and Placerville localities. Burro clays yield highest vanadium recoveries, followed by Deremo and Placerville clays in decreasing order of amenability.

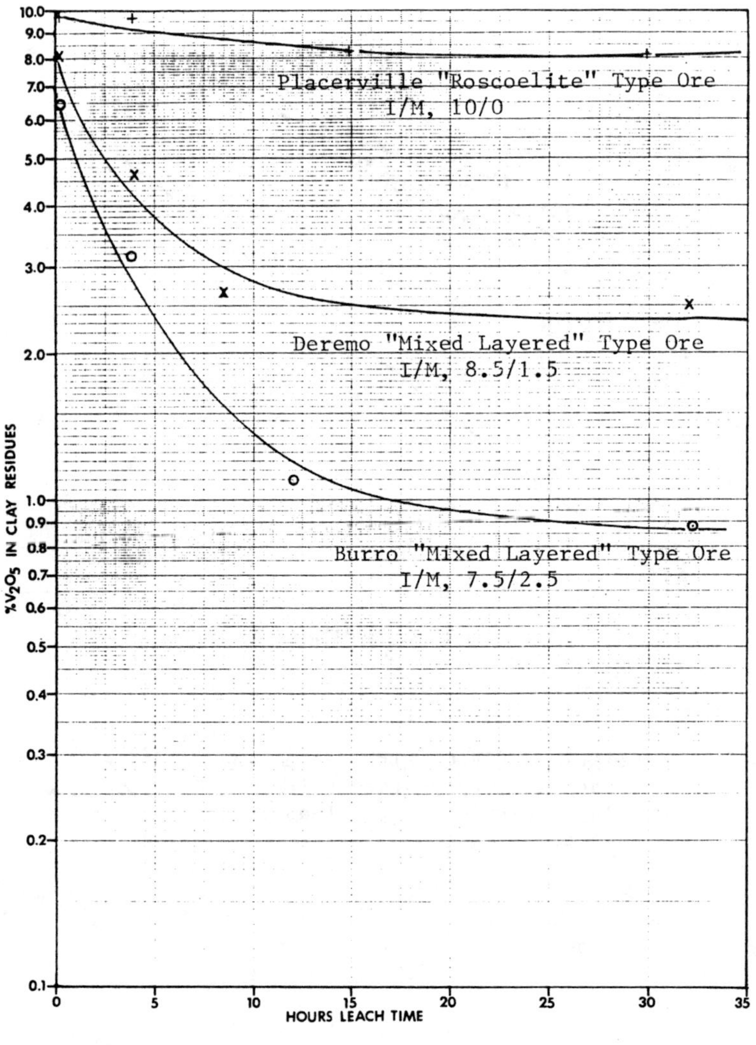

Fig. 9. Percent V_2O_5 in clay tails after varying leach times.

Burro clays also contain the highest percentage of expanding layers, followed by Deremo and finally Placerville clays, which do not contain measurable amounts of expanding clays. A direct correlation of expanding layers with vanadium extractions is apparent after 32 hours of leaching (Fig. 10). Measurement of mixed layering is concluded to be a potential guide to the amenability of certain clays to acid treatment.

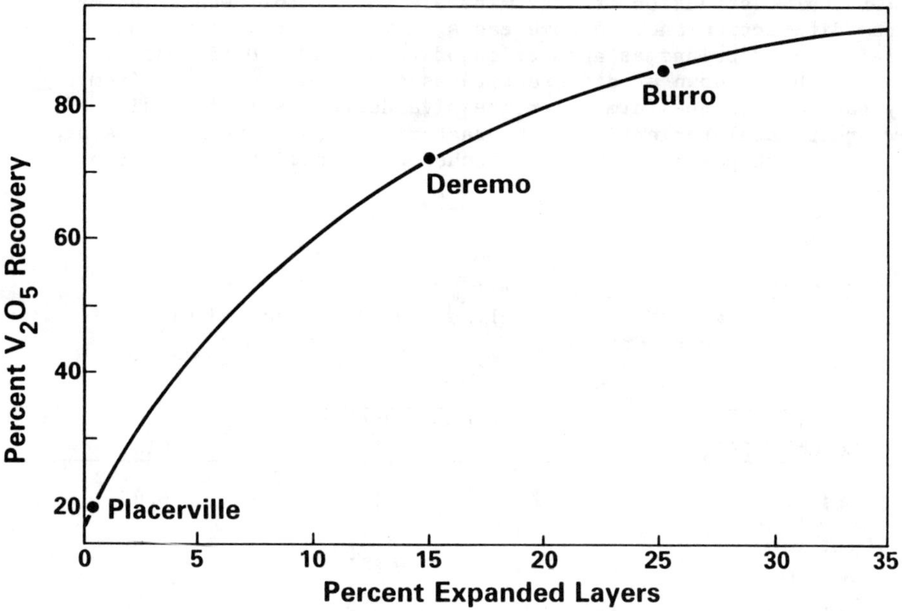

Fig. 10. Correlation of V_2O_5 recoveries with percent expanded layers in clays from Burro, Deremo, and Placerville samples.

It was interesting to note that the solubilization of vanadium from mixed layered clays had a relatively constant rate up to about ten hours, after which a rapid decrease in rate of solution occurred (Figure 9). A relatively flat curve is produced beyond this point of inflection which may be called the baseline limit of amenability. Expanding layers of montmorillonite seem to permit limited access of leach solutions between the clay layers, and to allow proportionate extraction of vanadium from available octahedral positions in the illite sheets.

BREAKDOWN OF CLAY STRUCTURES BY ACID TREATMENT

Very little has been reported in available literature regarding such parameters as (1) clay mineralogy of different ores, (2) vanadium content of different clays, and (3) the relative solubility of vanadium from different clay minerals. It now becomes more apparent that clay mineralogy is a significant variable, equal in importance to temperature, pH and contact time in hot acid leaching. Without pertinent data on mineralogic parameters, evaluations of amenability can remain at best only empirical.

Gradual destruction of clay minerals can be followed directly by X-ray diffraction and, in some cases, may be correlated with vanadium extractions as a function of treatment. Different clay minerals break down at different rates that are measurable from X-ray patterns in the form of progressive decreases in intensity of low order basal reflections. Characteristic (00ℓ) reflections are compared with percent V_2O_5 for leached clay residues of Burro ore in Table III.

TABLE III. A Comparison of I/I_o Values of (00ℓ) Reflection with Vanadium Analyses for Acid Leached Clays of Burro Ore

Contact Time (Acid Leach)	Chlorite I/I_o (001)	Chlorite (004)	Illite-Montmorillonites I/I_o (001)	$\%V_2O_5$ in Clay Residue
0 hrs	100%	100%	100%	6.85
4 hrs	50	35	85	3.25
12 hrs	25	15	65	1.08
32 hrs	<10	<10	50	<1.00 (0.89)*

* 0.89 value obtained with x-ray fluorescence.

A compilation of data from over 60 x-ray patterns of acid-leached clays indicates that vanadiferous chlorites are more easily broken down by acid treatment than mica-montmorillonites. Basal reflections are reduced significantly in intensity after four hours of acid treatment, and are largely destroyed after 12 hours. The chlorite lattice appears to be largely destroyed and the vanadium recovered during early stages of acid treatment.

Vanadiferous illite, and mixed layered mica-montmorillonites are more acid resistant than chlorite, and remain relatively crystalline even after 30 or more hours of acid leaching at elevated temperatures ($80^{o}C$). Illites that contain no interstratified montmorillonite are the most refractory to acid leaching of any of the layered silicates, and retain the highest degree of crystallinity after comparable leaching conditions.

THE MECHANICS OF ACID DECOMPOSITION OF MIXED LAYERED CLAYS

X-ray diffraction and analytical data on leached clay residues from several ores have indicated that the presence of interstratified montmorillonite between illite layers improves the recovery of vanadium by acid leaching. Earlier studies by P. G. Nutting (Nutting, 1943) on the solubilities of clay minerals of the montmorillonite group explain why montmorillonites are beneficial in mixed layered clays. Dr. Nutting's studies reveal that silicates of the montmorillonite group are much more easily attacked by acids than those of the kaolin, mica, talc, or pyrophyllite groups. The bases, Al_2O_3, Fe_2O_3, MgO, CaO, etc., are readily removed from montmorillonite in a few hours by hot 6 N HCl, leaving an amorphous residue of silica.

Evidence has been assembled which indicates that montmorillinitic layers serve as passageways for leach solutions in the extraction of vanadium from mica-montmorillinites. Stages in the removal of vanadium from mixed layered clays by acid leaching are outlined below:

(1) Expansion of interstratified layers of montmorillonite by aqueous leach solutions.

(2) Penetration of leach solutions along expanded montmorillonite layers between illite sheets.

(3) Acid attack and solubilization of R_2O_3 and RO components from the montmorillonite.

(4) Lateral penetration of acid solution into adjacent illite sheets.

(5) Solubilization and removal of V_2O_5 from available octahedral positions in the illite sheets.

(6) Decline in rate of removal of vanadium as rate of acid penetration into adjoining illite sheets decreases.

The decline in rate of removal of vanadium from a given clay after about 10 to 15 hours of leaching is shown by leaching curves in Figure 9 and corresponds to the baseline limit of amenability. This limit of amenability to acid leaching is a measurable parameter, distinct for any mixed layered clay, and depends on the frequency and proximity of expanding layers to the vanadium-bearing illite sheets.

In vanadiferous clay micas with little or no mixed layering, the baseline limit of amenability is relatively high, and methods of beneficiation based on hot acid leaching would be uneconomic. The effectiveness of acid treatment depends largely on the relative abundance of expanding to non-expanding layers in clay micas, which in some ores pose definite limitations that probably must be overcome by roasting at elevated temperatures above 800°C, instead of acid leaching.

GENERAL EFFECTS OF CLAY MINERALOGY ON VANADIUM RECOVERY

The significance of clay mineralogy to the amenability of vanadium in sandstone ores has been described above, in which the degree of mixed layering of mica-montmorillonites was related to vanadium extractions. Much of the vanadium in sandstone ores is tied up in various clay silicates that require variable types of treatment for optimum recovery. Most of the clays in the additional eighteen metallurgical pulps listed with amenability data in Table IV are in the form of clay micas (illite) and mixed layered mica-montmorillonites. Other clay minerals include kaolinite, chlorite, and various chlorite-montmorillonites and illites.

Samples with higher clay content generally have higher proportions of the vanadium combined in the clays. Vanadium is extracted with difficulty from clay silicates by acid leaching, and consequently, samples of higher clay content usually give poorer extractions. Although there are other significant variables to consider, only the total clay content of amenability pulps are plotted against vanadium extractions in Figure 11. Samples of equivalent vanadium grade (1.0-2.0% V_2O_5) are usually more amenable to acid leaching when little clay is present because of less vanadium combined in clays and more in the form of soluble oxides.

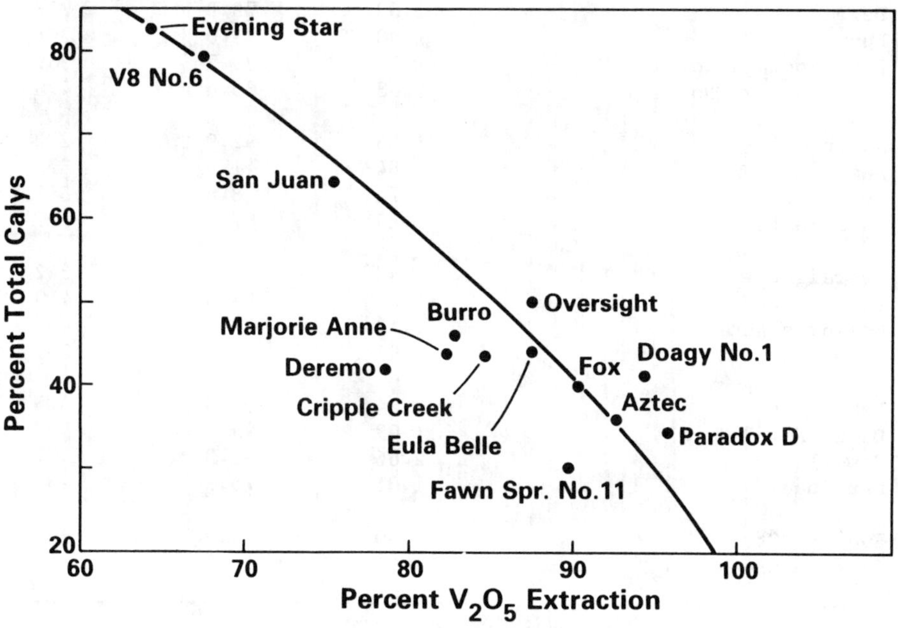

Fig. 11. Correlation of V_2O_5 extractions with clay contents for samples of equivalent V_2O_5 content (1.0-2.0% V_2O_5).

TABLE IV. Amenability Data on Sample Pulps from Uravan*

Claim	% V_2O_5 (Total)	% V_2O_5 Extracted	lbs Acid/ Ton
Deremo	1.63	78.8	278
Burro	1.40	83.0	391
Fawn Springs No. 11	1.12	89.7	369
Fawn Springs No. 18	2.68	83.4	498
Aztec	1.08	92.9	230
Sam	0.84	41.9	378
Zee Lease	0.72	78.8	317
Eula Belle	1.25	87.8	299
Oversight	1.43	87.8	342
Fox	1.86	90.6	410
Marjorie Anne	1.12	82.3	402
V8 No. 6	0.97	67.7	291
Paradox D	1.24	95.3	325
Doagy No. 2	1.09	94.7	307
Capella	2.66	93.3	477
San Juan	1.01	75.7	257
Evening Star	1.50	64.2	356
Cripple Creek No. 2	1.64	84.9	305

* According to the amenability procedure used at Uravan, 1000 grams of minus 14 mesh ore were acid leached with agitation at 50 percent solids for 18 hours at a maintained temperature of 80-82°C. An initial addition of acid was made to bring the pH of the slurry to 0.4 pH, where it was maintained for 8 hours with additional acid.

Mixed layering of mica-montmorillonites has been established as a significant factor in the acid amenability of sample pulps of relatively high clay content. Mica-montmorillonites of higher montmorillonite content are usually more easily extracted than those of lower content. Measurements of mixed layering for samples of high equivalent clay content are plotted in Figure 12 to illustrate the effects of expanded layers of montmorillonite on vanadium extractions. Although there are insufficient numbers of samples of equivalent high vanadium content (40 percent clay) in Figure 12 to establish valid curves, the higher recovery of vanadium from clays of higher expanded layering is, nevertheless, apparent. Measurements of mixed layering are tabulated with (1) total clay, (2) total V_2O_5, and (3) V_2O_5 extractions in Table V.

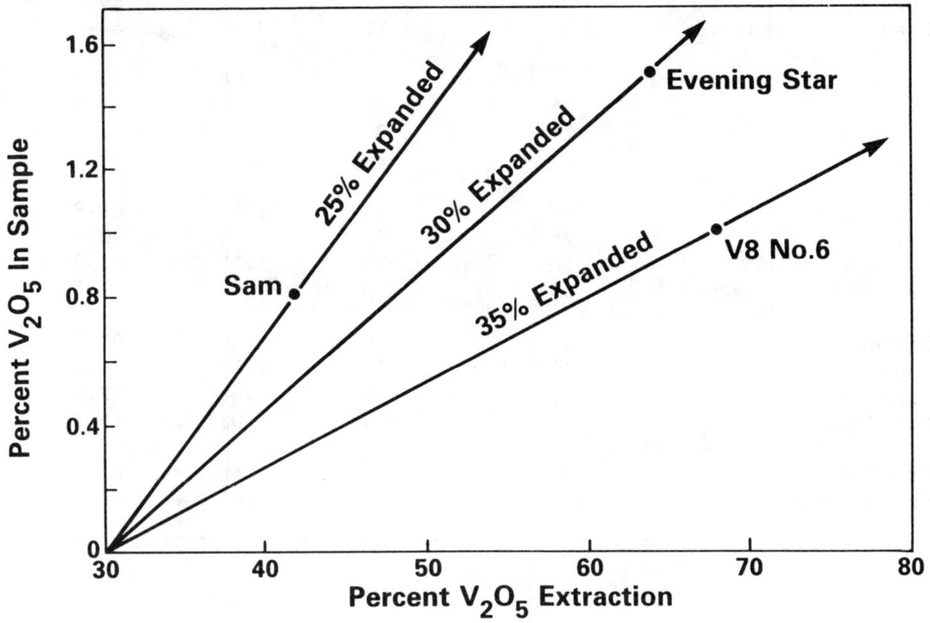

Fig. 12. Effect of expanded layers of montmorillonite on vanadium extractions for sample of equivalent clay content (40% clays).

TABLE V. Mineralogical and Metallurgical Factors Significant to Vanadium Recoveries (Semiquantitative Wt %)

Claim	% Expanded Layers in Mica-Montmorillonites	% Total Clays	% V_2O_5 (Total)	% V_2O_5 Extraction
Deremo	20	20	1.63	78.8
Burro	25	25	1.40	83.0
Fawn Spring No. 11	30	15	1.12	89.7
Fawn Spring No. 18	30	25	2.68	83.4
Aztec	35	20	1.08	92.9
Sam	25	40	0.84	41.9
Zee Lease	25	30	0.72	78.8
Eula Belle	25	20	1.25	87.8
Oversight	30	25	1.43	87.8
Fox	25	20	1.86	90.6
Marjorie Anne	30	20	1.12	82.3
V8 No. 6	35	40	0.97	67.7
Paradox D	20	15	1.24	95.3
Doagy No. 2	30	20	1.09	94.7
Capella	25	25	2.66	93.3
San Juan	30	30	1.01	75.7
Evening Star	30	40	1.50	64.2
Cripple Creek	25	20	1.64	84.9

The amenability of samples containing less than 20 percent clays seems to be little affected by variations in clay mineralogy. A marked decrease in the effect of mixed layering on vanadium extractions is illustrated in Figure 13, where vanadium extractions are plotted against total clay content as a function of percent expanded layers in the mica-montmorillonites. Maximum divergence in vanadium extractions is apparent in samples of highest clay content. Marked lessening of this divergence occurs in samples of lower clay content until variations in extractions are minimized in samples of less than 20 percent clay content.

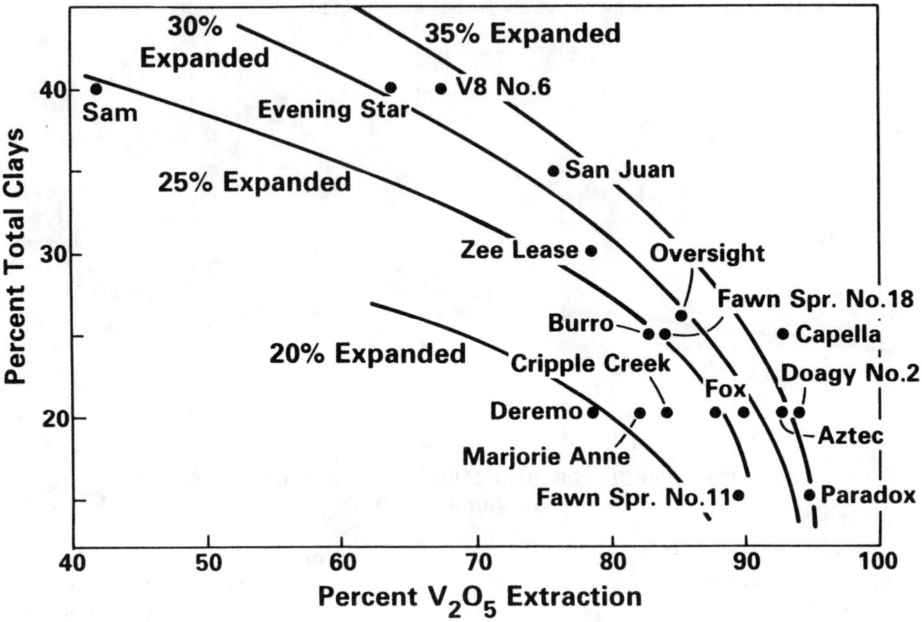

Fig. 13. Correlation of V_2O_5 extraction with clay content as a function of mixed layering.

Only four factors have been considered in the evaluation of vanadium recoveries in the current study: (1) total clays, (2) total vanadium, (3) percent expanded layers, and (4) percent vanadium extracted. A number of other factors have not been considered in this study, some of which include: (1) percent vanadium in the clays, (2) relative distribution of vanadium in other clay minerals, chlorite, illite, etc., (3) oxidation state of the ore, (4) amounts of vanadium in reduced and oxidized oxides and vanadates, etc. Some of these factors can be determined with difficulty, if at all. Consequently, only the four factors believed to be most significant in the majority of amenability samples were selected for this investigation. These factors are readily determined by x-ray diffraction and wet analyses, and simply illustrated in the form of graphs (Figs. 12 and 13).

The data indicate that these four variables may be used in a limited sense to estimate some of the amenability characteristics of vanadium ores, e.g., percent V_2O_5 extractions and acid consumption. An attempt is made in the following section to correlate mineralogical parameters with acid consumption.

THE EFFECTS OF MINERALOGY ON ACID CONSUMPTION OF VANADIUM ORES

The effects of mineralogy on the acid consumption of vanadium ores were evaluated on the basis of amenability and mineralogical data listed in Tables IV and VI. The acid-consuming components of the ore were integrated into an acid consumption factor (A.C.F.) which correlates closely with the acid consumption figures in Table IV. The acid consumption factor was arbitrarily computed for each ore sample from the following formula:

$$\text{A.C.F.} = 1/3(\text{Total \% Clays})(\text{\%Expanded Layers}) + \text{Total \% Carbonates} + 2\ (\text{Total \% } V_2O_5)$$

Montmorillonite (expanded) layers, carbonates, and vanadium minerals are known to be acid consumers in most ore feed, and are empirically integrated into the above formula to obtain an overall factor that may be used to express acid consumption. Acid consumption factors are listed in Table VII with the pounds of sulfuric acid used per ton of ore. A plot of values in Figure 14 indicates that acid consumption values may usually be estimated from mineralogic data to within 70 pounds of acid per ton of ore. Sample V8 No. 6 is noticeably out of line with the rest of the data. Reasons for this departure are not known.

TABLE VI. Mineralogy of Sample Pulps
(Semiquantitative Wt %)

Claim	Qtz	Feld-spar	Cal-cite	Dolo-mite	Gypsum	% Clays (Remain.)	Total %
Deremo	78	1	<1	<1	0	21	100
Burro	71	2	3	0	0	23	100
Fawn Springs No.11	72	8	<1	5	0	25	100
Fawn Springs No.18	64	7	0	4	0	25	100
Aztec	71	10	<1	1	0	18	100
Sam	52	3	5	1	0	39	100
Zee Lease	59	7	5	<1	0	29	100
Eula Belle	69	7	0	2	0	22	100
Oversight	55	6	3	1	10	25	100
Fox	70	7	0	3	0	20	100
Marjorie Anne	67	6	0	5	0	22	100
V8 No. 6	48	7	6	0	0	39	100
Paradox D	48	5	3	2	25	17	100
Doagy No. 2	70	7	2	<1	0	21	100
Capella	63	7	0	4	0	26	100
San Juan	59	6	3	0	0	32	100
Evening Star	48	7	0	3	0	42	100
Cripple Creek No.2	66	11	0	1	0	22	100

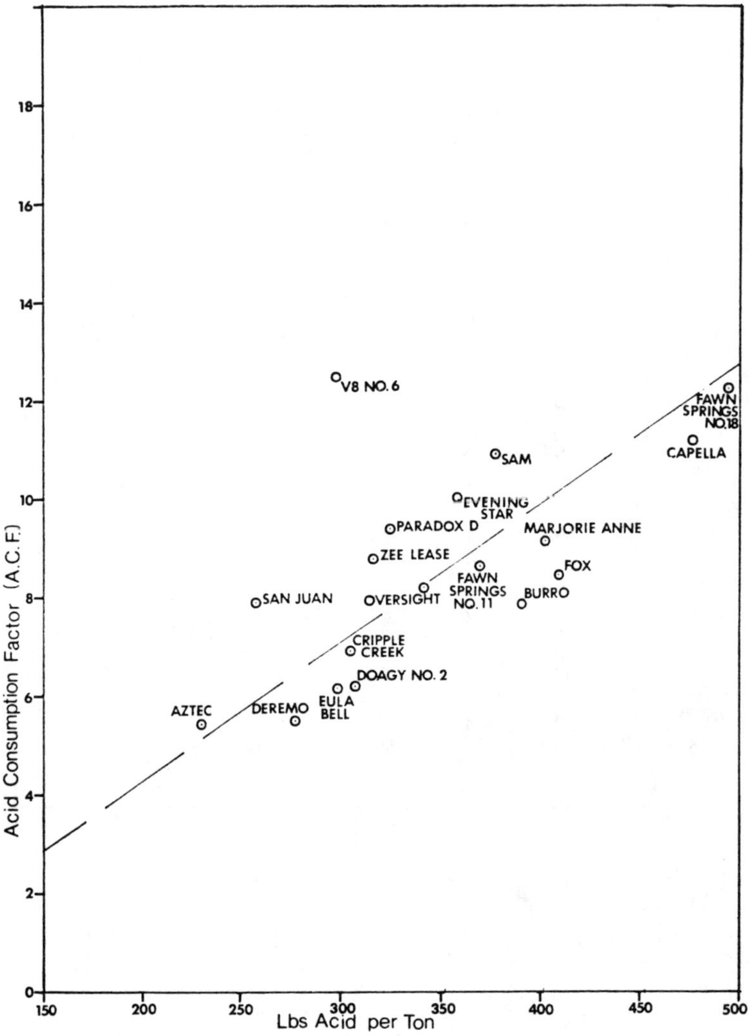

Fig. 14. Correlation of acid consumption factor (A.C.F.) with lbs acid per ton of ore.

TABLE VII. Acid Consumption Data

Claim	A.C.F.	Lbs Acid/Ton
Deremo	5.6	278
Burro	7.9	391
Fawn Springs No. 11	8.7	369
Fawn Springs No. 18	12.3	498
Aztec	5.5	230
Sam	11.0	378
Zee Lease	8.9	317
Eula Belle	6.2	299
Oversight	8.3	342
Fox	8.5	410
Marjorie Anne	9.2	402
V8 No. 6	12.6	291
Paradox D	9.5	325
Doagy No. 2	6.2	307
Capella	11.4	477
San Juan	8.0	257
Evening Star	10.0	356
Cripple Creek No. 2	7.0	305

REFERENCES

Bradley, W. C., 1945, "Diagnostic Criteria for Clay Minerals," Am. Mineral., Vol. 30, pp 704-713.

Bradley, W. C., 1953, "Analysis of Mixed-Layer Clay Mineral Structures," Analytical Chem., Vol. 25, p 727.

Brown, G. and MacEwan, D. M. C., 1950, "The Interpretation of X-Ray Diagrams of Soil Clays, Part II," J. Soil Sci., Vol. 1, pp 239-253.

Brindley, G. W., 1951, "X-Ray Identification and Crystal Structures of Clay Minerals," Mineralogical Society of Great Britain.

Foster, Margaret D., 1959, "Chemical Study of the Mineralized Clays," U.S.G.S. Prof. Paper 320, pp 121-132.

Grim. R. E., Bradley, W. F., and Brown, G., 1951, "X-Ray Identification and Crystal Structures of Clay Minerals," Mineralogical Society, London, Chap. V, pp 138-172.

Hathaway, John C., 1955, "Mixed Layered Structures in Vanadium Clays, Clays and Clay Minerals," Publication 456, Abstr.

Hathaway, John C., 1959, "Mixed Layered Structures in Vanadium Clays, Clays and Clay Minerals," U.S.G.S. Prof. Paper 320, pp 133-138.

Hausen, D. M., 1983, "Process Mineralogy of the $Na_2O-V_2O_5$ System in Vanadium Production," SME-TMS-AIME Process Mineralogy Symposium, Atlanta, GA.

Hendricks, S. B. and Teller, E., 1942, "X-Ray Interference in Partially Ordered Layer Lattices," J. Chem. Phys., Vol. 10, pp 147-167.

MacEwan, D. M., 1942, "Some Notes on the Recording and Interpretation of X-Ray Diagrams of Soil Clays," J. Chem. Phys., Vol. 10, pp 147-167.

Nutting, P. G., 1943, "The Action of Some Aqueous Solutions on Clays of the Montmorillonite Group," U.S.G.S. Prof. Paper 197.

Weaver, Charles E., 1955, "The Distribution and Identification of Mixed-Layer Clays in Sedimentary Rocks," Clays and Clay Minerals, Publication 456, Abstr.

Weaver, Charles E., 1956, "The Distribution and Identification of Mixed-Layer Clays in Sedimentary Rocks," Am. Mineral., Vol. 41, pp 202-221.

Chapter 9

CLAY MINERALOGY OF INSOLUBLE RESIDUES IN MARINE EVAPORITES

Marc W. Bodine, Jr.

U.S. Geological Survey
MS 939, Denver Federal Center, P. O. Box 25046
Denver, Colorado 80225

ABSTRACT

Insoluble residues from three sequences of Paleozoic marine evaporites (Retsof salt bed in western New York, Salado Formation in southeastern New Mexico, and Paradox Member of the Hermosa Formation in southeastern Utah) are rich in trioctahedral clays. Chlorite (clinochlore), corrensite (mixed-layer chlorite-trioctahedral smectite), talc, and illite (the only dioctahedral clay) are the dominant clay minerals; serpentine, discrete trioctahedral smectite (saponite), and interstratified talc-trioctahedral smectite are sporadically abundant. These clay-mineral assemblages differ chemically and mineralogically from those observed in most continental and normal marine rocks, which commonly contain kaolinite, dioctahedral smectite (beidellite-montmorillonite), illite, mixed-layer illite-dioctahedral smectite, and, in most cases, no more than minor quantities of trioctahedral clay minerals. The distinctive clay mineralogy in these evaporite sequences suggests a largely authigenic origin. These clay minerals are thought to have formed during deposition and early diagenesis through interaction between argillaceous detritus and Mg-rich marine evaporite brines.

INTRODUCTION

The insoluble mineral fraction in marine evaporite rocks consists chiefly of quartz and clay minerals along with variable proportions of feldspar, hematite, pyrite, and other silicate and oxide resistates (Stewart, 1963; Braitsch, 1973). The mineralogy and geochemistry of these minerals, for the most part, have received considerably less attention than that of their saline hosts. Yet, their occurrence has long had significant geologic and commercial importance.

The silicates and oxides are disseminated ubiquitously throughout saline rocks in highly variable quantities. They frequently tend to be differentially concentrated along bedding planes, and in some cases (see Dellwig and Evans, 1969, for example), are the only criteria in the rock for evaluating local bedding and small-scale geologic structures. The silicate fraction is also sporadically heavily concentrated in distinct lenses, pods, and angular-to-rounded lithic fragments of varying size and abundance (see Mayrhofer, 1955, for example), and as such yield valuable information on the sedimentologic character of the depositional environment and salt genesis. Furthermore, clastic seams or beds (sand and claystones), which range from a few millimeters to tens of meters in thickness, are commonly intercalated with salt beds; these units not only allow for more precise stratigraphic correlation, but frequently provide additional sedimentologic criteria for interpreting the depositional history of the saline section.

Variations in the character and abundance of the insoluble fraction throughout a salt section are of importance to salt producers. Dry mining operations are commonly influenced by the occurrence and character of clastic interbeds; the clay abundance in salts may well affect drilling parameters for solution mining (see Saberian, 1980, for example); and the abundance and character of disseminated clays in recovered salts may affect profit and impose constraints on milling and refining operations (see, Case, 1978, for example). Many of these concerns apply to the use of salt caverns for resource storage and to recent investigations evaluating the suitability of saline strata as potential hosts for permanent storage of nuclear waste. For the latter purpose, an understanding of the chemical and mineralogic response of the salt's included silicates to imposed thermal perturbations in a saturated brine environment becomes important.

This paper focuses on the character of clay minerals in insoluble fractions from three marine evaporite deposits. These clays are mineralogically and chemically highly distinctive. Although often referred to as part of the "detrital fraction" of saline rocks, the clays, for the most part, are authigenic. They appear to have formed in response to the saline chemistry in the depositional and subsequent diagenetic environments and occur both as replacement products of original argillaceous detritus, and as neoformed silicates that precipitated directly from the brine.

Evaporite localities

The clay mineral assemblages are from three Paleozoic marine salt deposits: Ochoan salts from the Carlsbad district in southeastern New Mexico; Paradox salts from the Paradox Basin in southeastern Utah; and the Retsof salt bed from the western margin of the Appalachian Basin in western New York State.

Ochoan (uppermost Permian) salts in the Delaware Basin of southeastern New Mexico and adjacent parts of Texas occur in two major formations: the Castile Formation (up to >1000 m thick), an alternation of bedded anhydrites and thick halite intervals deposited in relatively deep-water (Dean, Davies, and Anderson, 1975); and the overlying Salado Formation (up to >600 m thick), a shallow-water succession of repeated halite-dominated cycles with each bounded by or containing one or more thinner beds of anhydrite, mudstone, polyhalite, or soluble potash salts (Cheeseman, 1978; Jones, 1954; Powers, 1978). Polyhalite replacement of many anhydrite beds, frequent occurrences of magnesite, unusual bromine profiles in the salts, and the Mg-depleted character of the soluble potassium salt assemblages suggest postdepositional fluid mobility with resultant salt recrystallization and solute transport. East of Carlsbad along the northern margin of the Delaware Basin, the McNutt potash zone in the Salado Formation contains the rich sylvite and langbeinite bearing beds that are the nation's most important domestic source of potash. Clays were analyzed from all lithologies in the Salado Formation and from uppermost anhydrites of the Castile Formation; the samples were from the ERDA-9 drill hole that supported geologic characterization studies preceeding construction of the Waste Isolation Pilot Plant (WIPP) on the southern perimeter of the potash district (Powers, 1978). Maximum depth of burial of Ochoan rocks in the Delaware Basin did not exceed a kilometer with maximum temperatures of <60°C.

Salts from the Paradox Member of the Hermosa Formation (Middle Pennsylvanian) reach a thickness in excess of 700 m in southeastern Utah. The saline section of the Paradox Member has been divided into a succession of 26 discrete cycles of salt deposition; with each cycle ideally illustrating recharge of the basin with normal seawater, followed by progressive evaporation with depositon of carbonate, then calcium sulfate, thick halite, and in several cycles, culminating with potash salt precipitation (Hite, 1960; Hite and Buckner, 1981). Bromine profiles through halite intervals from most cycles suggest that the halite is unaltered and was deposited during progressive evaporation of seawater (Raup, Hite, and Groves, 1970). Clay samples were obtained throughout the Paradox Member from the Gibson Dome No. 1 drill core (Hite, 1983), and detailed samples from a single depositional cycle from the Cane Creek No. 1 drill core (Raup, 1966). Maximum depth of burial of the Paradox salts was probably 2-3 kilometers with maximum temperatures of 60-80°C if not influenced by nearby intrusives.

Salt from the Retsof Mine (International Salt Co.) is produced from a 3-4 m-thick halite bed, the Retsof salt bed, in the Vernon Formation (Upper Silurian). The formation is the basal unit in the Appalachian Basin section of the extensive Salina Group (Rickard, 1969). Sedimentologic evidence and bromine distribution profiles through the halite at Retsof (Dellwig and Evans, 1969) strongly suggest a shallow (above wave base), subaqueous, hypersaline environment of marine origin. Clay samples were obtained principally from the Retsof salt bed, as well as from fragmental, commonly rounded, dolomite inclusions in the salt, and from the bedded dolomites both immediately above and below the salt bed. Maximum depth of burial of the Retsof salt bed likely produced temperatures of about 100° based on conodont maturation data (Harris, Harris, and Epstein, 1978).

CLAY-MINERAL ASSEMBLAGES

Clay-mineral assemblages and associations discussed in this paper utilize both published (Bodine and Standaert, 1977; Bodine, 1978; Bodine and Rueger, 1984) and unpublished data from each of the three localities. Table 1 lists the clay minerals observed in these sections, and table 2 gives the chemical composition for nine representative clay-mineral assemblages.

Clay separation and analytical procedures

Separation of the clay fraction involved dissolution of the water-soluble salts with a water leach, and dissolution of the "acid-soluble" salts (Ca-Mg carbonates and Ca sulfates) by cation-complexing with Na_4-EDTA (versane) in a slightly alkaline solution (Bodine and Fernalld, 1973). This was followed by particle disaggregation, if necessary, with gentle grinding and particle dispersion in distilled water using an immersible ultrasonic probe. Cycles of centrifuging, decanting off the supernatant solution, and redispersing the particulates by sonification in distilled water were repeated until flocculation was eliminated. Size-fractions of the dispersed particulates were separated with timed, speed-controlled centrifugations. A variety of size-fractions were separated with the <2 or <1 μm effective spherical diameter fractions effectively removing most non-clay constituents.

Clay determinative procedures relied chiefly on X-ray diffraction. Diffractometer scans to determine the succession of basal spacings were from oriented sample mounts, i.e., mounts prepared with clay flakes dispersed in water and sedimented onto glass slides, then air-dried. Each oriented sample was analyzed untreated (air-dried), saturated with ethylene glycol, and after heating to 350 and 550°C. Dried samples of the clay-size fractions, packed in conventional diffractometer sample holders, were analyzed to determine the general array of hkl spacings for each assemblage. The cumulative diffractometer data were evaluated using recent criteria for clay mineral identification (Brown and Brindley, 1980) and evaluating extent and character of interstratification in mixed-layer clays (Reynolds, 1980). Many of the clay fractions were chemically analyzed for major rock-forming components by either rapid silicate analysis techniques with atomic absorption spectrometry, or by quantitative X-ray fluorescence techniques (Taggert, Lichte, and Wahlberg, 1981).

Clay-mineral characteristics

Illite is the only dioctahedral clay in these rocks and displays characteristics which persist throughout the three localities. The illite is remarkably well crystallized; it produces sharp diffraction maxima that are more closely akin to a muscovite or sericite than to a

TABLE 1. Clay minerals in marine evaporites (the Ochoan Series, the Paradox Member of the Hermosa Formation, and the Retsof salt bed,); their chemical formula, and formula source; mineral listing in order of increasing Mg/Al atom ratio

Clay mineral	Formula	Source*
Illite	$K_{0.85}(Al_{1.61}Fe^{3+}_{0.20}Mg_{0.23})(Al_{0.76}Si_{3.24})O_{10}(OH)_2$	1
	$(K_{0.72}Na_{0.03}Ca_{0.03})(Al_{1.38}Fe^{3+}_{0.38}Mg_{0.25})(Si_{3.40}Al_{0.60})O_{10}(OH)_2$	2
	$K_{0.8}(Al_{1.69}Mg_{0.35})(Si_{3.43}Al_{0.57})O_{10}(OH)_2$	3
Chlorite (clinochlore)	$(Mg_{4.51}Fe^{2+}_{0.23}Al_{1.20})(Al_{1.09}Si_{2.91})O_{10}(OH)_8$	1
	$(Mg_5Al)(Si_3Al)O_{10}(OH)_8$	4
Saponite	$Mg_{0.2}(Mg_3)(Si_{3.6}Al_{0.4})O_{10}(OH)_2$	4
Interstratified chlorite-saponite	$[(Mg_5Al)(Si_3Al)O_{10}(OH)_8]_x[Mg_{0.2}(Mg_3)(Si_{3.6}Al_{0.4})O_{10}(OH)_2]_{1-x}$	4
Corrensite	$(Na_{0.26}K_{0.02})(Mg_{7.59}Fe^{2+}_{0.25}Ti_{0.02}Al_{0.95})(Si_{7.12}Al_{0.88})O_{20}(OH)_{10}$	5
	$Mg_{0.2}(Mg_8Al)(Si_{6.6}Al_{1.4})O_{20}(OH)_{10}$#	4
Interstratified talc-saponite	$[Mg_3Si_4O_{10}(OH)_2]_x[Mg_{0.2}Mg_3(Si_{3.6}Al_{0.4})O_{10}(OH)_2]_{1-x}$	4
Talc	$Mg_3Si_4O_{10}(OH)_2$	4
Serpentine	$Mg_3Si_2O_5(OH)_4$	4

*Source:
1. Mean composition in Retsof salts (Bodine and Standaert, 1977).
2. Silver Hill illite (Srodoń and Eberl, in press).
3. Idealized sedimentary illite (Perry, 1971).
4. Idealized stoichiometry; smectites assigned layer charge deficiency of 0.4 to facilitate calculation of idealized clay reactions.
5. Analysis of nearly monomineralic clay fraction from the Paradox section (Bodine and Rueger, 1984).

#Identical to interstratified chlorite-saponite (above) with x = 0.5.

TABLE 2. Chemical analyses of representative clay-mineral assemblages in marine evaporites

[Total Fe as Fe_2O_3; LOI--loss on ignition, 900°C; nd--not determined]

	1	2	3	4	5	6	7	8	9
SiO_2	55.9	41.2	39.6	60.4	41.4	42.9	46.7	45.2	54.4
TiO_2	0.66	0.33	0.27	0.81	0.37	0.15	0.26	0.43	0.20
Al_2O_3	17.32	19.4	15.8	14.30	15.9	9.31	7.97	6.56	3.80
Fe_2O_3	5.75	3.61	4.44	5.29	3.92	2.00	6.84	12.18	3.76
MnO	0.07	<0.02	0.02	0.07	<0.02	<0.02	0.05	0.05	nd
MgO	4.32	16.3	23.7	5.67	22.8	30.7	21.4	15.6	27.7
CaO	0.12	<0.02	0.35	0.14	<0.02	<0.02	0.12	0.32	0.09
Na_2O	0.20	0.53	0.43	0.30	0.87	0.79	0.18	0.34	0.35
K_2O	6.66	4.07	2.07	3.82	2.01	0.11	0.45	0.89	0.56
P_2O_5	0.05	0.06	0.16	0.05	<0.05	<0.05	0.03	0.05	nd
S	1.87	nd	nd	1.10	nd	nd	3.13	6.73	nd
LOI	nd	13.2	12.1	nd	12.0	13.2	nd	nd	nd

1. Illite-rich with minor clinochlore-pyrite, Retsof halite (unpub.).
2. Illite-rich with interstratified chlorite-saponite, Paradox sylvite-halite (unpub.).
3. Chlorite-rich with illite, Paradox dolomite (unpub.)
4. Chlorite-illite-pyrite, Retsof halite (Bodine, in press).
5. Corrensite-rich with illite, Paradox anhydrite (unpub.).
6. Corrensite, Paradox carnallitic halite (Bodine and Rueger, 1984).
7. Talc-rich with chlorite-pyrite-illite, Retsof halite (Bodine, in press).
8. Talc-serpentine-pyrite-illite-chlorite, Retsof halite (unpub.)
9. Interstratified talc-saponite with minor illite, Salado polyhalite, Delaware Basin, (Bodine, in press).

sedimentary illite. The illite characteristically contains only negligible, if any interstratified smectite--again a rarity in sedimentary illites (Reynolds, 1980). Microprobe analyses of Retsof clays (Bodine and Standaert, 1977) document the illite's mica-like, smectite-free character; its mean composition (table 1) shows more tetrahedral Al and more interlayer cations per formula unit than reported for most analyzed (Srodoń and Eberl, in press; Weaver and Pollard, 1973) or idealized (Perry, 1971) sedimentary illite compositions.

Chlorite is equally distinctive. Chemical analyses of chlorite-rich assemblages from the three localities (table 2), and the mean structural formula (table 1) from microprobe analyses of the Retsof chlorites (Bodine and Standaert, 1977) clearly document the clinochlore composition (Foster, 1962) and the unusually high Mg/Fe+Mg mole fraction in these chlorites (0.95 in the mean Retsof chlorite). The same mole fraction in normal sedimentary clay chlorites, on the other hand, rarely exceeds 0.7 and is often less than 0.5 (Weaver and Pollard, 1973).

Chlorite occurs as a discrete clay mineral at the three localities, and at two (the Delaware and Paradox Basins), it also occurs as a component in mixed-layer clays interstratified with trioctahedral smectite. Based on chemical analyses of the assemblages with abundant mixed-layer clay (table 2), the chlorite layers in the interstratified clays are also substantially enriched in magnesium.

Relative abundances of layer types in the interstratified chlorite-smectite lie along the binary continuum from chlorite to trioctahedral smectite (saponite). The most abundant interstratified chlorite-saponite composition is the 1:1, or near 1:1 chlorite-saponite. Particularly in the Paradox assemblages, the layer alternation pattern is frequently ordered and characterizes the clay as a corrensite (Bailey, 1982). Chemical analysis of corrensite from a nearly monomineralic Paradox assemblage yields a structural formula (table 1) that, except for a modest deficiency of octahedral cations, is very close to the ideal 1:1 interstratified clinochlore-saponite stoichiometry. Frequently samples with ~50% smectite layers are more randomly interstratified and not corrensite. Excluding the ~1:1 chlorite-saponite composition, the chlorite-rich compositions are more common than the saponite-rich varieties, and discrete chlorite is far more abundant than discrete saponite. The latter has been observed in only few samples from the Delaware Basin section.

A second variety of interstratified clay, observed associated with polyhalite in the Salado Formation (Delaware Basin section), is mixed-layered talc-saponite; it appears to be disordered and its diffraction data suggest a layer composition of approximately 1:1.

Clay-mineral associations

The occurrence and association of the clay minerals thoughout the three localities are summarized in table 3. A number of relationships do appear that are of importance in discussing the origin of these assemblages.

Illite is ubiquitous and occurs throughout each locality, although its relative abundance in the clay-mineral assemblages varies markedly. Based on the range of K_2O values in chemical analyses of clay-mineral assemblages, illite abundances range from <10 to >70 weight percent.

Chlorite and interstratified chlorite-saponite are commonly the dominant trioctahedral clays, and, because of their closely related compositions and associations in these strata, they will be discussed together. With the exception of rare occurrences of discrete saponite,

TABLE 3. Summary of clay-mineral occurrences and associations in marine evaporite sections at Retsof, New York; the Paradox Basin, Utah; and the Delaware Basin, New Mexico

Illite	Occurs ubiquitously throughout sections at each locality; most commonly coexisting with discrete chlorite or interstratified chlorite-saponite.
Chlorite	Occurs ubiquitously at Retsof; abundant in basal penesaline intervals ("shale" and lower overlying dolomite) of Paradox salt cycles, but absent or minor in overlying interval of cycles; sporadically occurs throughout Delaware Basin section coexisting with interstratified chlorite-saponites.
Interstratified chlorite-saponite and corrensite	Absent from Retsof section; partially-to-well ordered, often corrensite, in hypersaline and upper penesaline intervals of Paradox salt cycles; generally poorly-to-partially ordered and abundant in Delaware Basin section, frequently smectite layers dominate, corrensite rare.
Saponite	Occurs sporadically in Delaware Basin section, most abundant in lower Salado and upper Castile Formations; absent from Retsof and Paradox Basin sections.
Talc	Occurs at each locality; abundant in Retsof and Paradox Basin sections where restricted to anhydrite and halite facies of progressive evaporation; equally restricted, but not common in Delaware Basin section.
Serpentine	Occurs sporadically in Retsof and Delaware Basin sections; restricted to halite immediately overlying anhydrite or dolomite (Retsof and Delaware Basin) or anhydrite immediately underlying halite (Delaware Basin); see Bodine (in press).
Interstratified talc-saponite	Occurs sporadically in Delaware Basin section; restricted to polyhalite beds and polyhalite-bearing anhydrite beds; absent from Retsof and Paradox Basin sections; see Bodine (in press).

these are the only Al-bearing trioctahedral species in the clay-mineral assemblages.

In the Retsof samples, only discrete chlorite is observed and it occurs ubiquitously--throughout the rock salts, the dolomite inclusions in the salt, and the bedded dolomites above and below the halite beds. Its composition, from microprobe analyses (Bodine and Standaert, 1977), remains remarkably constant throughout the Retsof section. No interstratified clays are found at Retsof.

In the Paradox Basin section, there is a pronounced tendency for discrete chlorite to be abundant in the basal penesaline beds (beds deposited from a seawater brine undersaturated with halite) that initiate each successive salt cycle, and interstratified chlorite-saponite is dominant in the halite beds of each overlying cycle, where it frequently occurs as corrensite (Bodine and Rueger, 1984). In Beth Madsen's and my current study of a single cycle from the Cane Creek No. 1 drill core, we are finding a clearly defined progression from discrete chlorite to 1:1 interstratified chlorite-saponite through the penesaline interval of the cycle (fig. 1). Discrete chlorite is abundant in the basal "shale" bed and immediately overlying dolomites that collectively mark initiation of deposition and progressive evaporation in the cycle. Above this horizon in the dolomite and overlying anhydrite, the chlorite contains increasingly more interstratified saponite layers, that approach the ordered 1:1 layer composition (corrensite) toward the top of the anhydrite interval.

Although there are indications of a parallel trend in the Ochoan salts, the data (Bodine, 1978) are far less convincing and often contradictory. Not only does the chlorite-saponite layer ratio in interstratified clay vary wildly and often unpredictably, but physical mixtures of discrete chlorite and smectite-rich chlorite-saponite are common and often preclude recognition of any systematic trends. Corrensite is not as abundant as in the Paradox salts.

Discrete saponite is not a common mineral in these assemblages. It is observed in a few samples from the Delaware Basin, and for the most part, is confined to the upper Castile and lower Salado anhydrites and rock salts.

Talc and serpentine are unusual sedimentary minerals, yet talc is frequently observed, occasionally in abundance, in clay-mineral assemblages at each locality. The talc is well crystallized and kerolite, the hydrous talc, was not observed. Serpentine, commonly coexisting with talc, has been unequivocally identified from Retsof and the Delaware Basin section, and its occurrence and abundance are considerably more restricted than talc's.

Talc occurs chiefly in the halite facies at the three localities, although it occasionally occurs in anhydrite and in assemblages with soluble potassium salts. Talc has not been observed in dolomites nor in the intercalated mudstones (Ochoan section) or basal "shale" beds

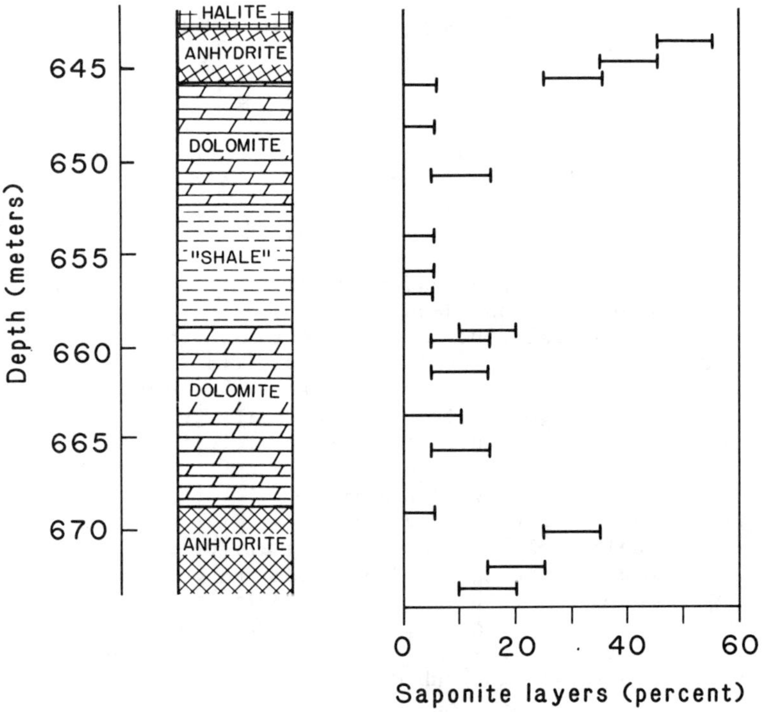

Fig. 1. Distribution of saponite-layer abundance in interstratified chlorite-saponites from penesaline beds at the base of salt 2 in the Paradox Member of the Hermosa Formation in Cane Creek No. 1 drill core, San Juan County, Utah. "Shale" is standard nomenclature for the Paradox Member; here the lithology is dolomitic argillaceous limestone.

(Paradox section). Talc is least abundant throughout the Ochoan section and reaches major proportions in some Retsof and Paradox assemblages.

Serpentine has been identified in some Retsof and Ochoan assemblages and is confined either to anhydrite beds immediately underlying bedded halite (Ochoan section), or to halites immediately overlying bedded anhydrite (Ochoan and Retsof sections).

Interstratified talc-saponite occurs only in the Salado Formation in the Delaware Basin and is confined to clay-mineral assemblages in polyhalite, or in one case, in an anhydrite bed containing substantial polyhalite. When it occurs, the interstratified talc-saponite is usually the dominant clay in the assemblage, and the normally abundant interstratified chlorite-saponite is either absent or its abundance is substantially reduced. Clay-mineral assemblages from polyhalite and anhydrite beds that contain no interstratified talc-saponite, consist of the characteristically abundant illite and interstratified chlorite-smectite with or without talc.

DISCUSSION

There appears little doubt that clay-mineral assemblages in marine evaporite rocks are enriched in magnesium, and trioctahedral clays tend to dominate the assemblages. Although the literature is not extensive, studies of clay minerals in other marine evaporite deposits illustrate several of these same features. For example, Füchtbauer and Goldschmidt (1959), in addition to documenting the pervasive occurrence of the chlorite-illite pair throughout the German Zechstein Formation, identified talc as a frequently occurring mineral in the Zechstein anhydrites; Niemann (1960) and Braitsch (1973) reported the restriction of corrensite presence to the potash salt and barren halite intervals, and talc's absence in the potash salt and basal mudstone (salzton) intervals in the Zechstein's Stassfurt Series.

The major conclusions from these three localities are: (1) the clays, for the most part, are authigenic; (2) the authigenic clays formed primarily in response to brine composition in the depositional and early diagenetic environments, and secondarily to later diagenetic and burial metamorphic effects; they crystallized either as replacements of the primary argillaceous clastic influx, or as direct precipitates from the coexisting brine; (3) some specific features of the clay-mineral assemblages, i.e., the distribution of talc and serpentine, and the association between interstratified talc-saponite and some polyhalites, can be attributed to specific events in the history of the saline section. The genesis of these latter minerals (talc, serpentine, and interstratified talc-saponite) are discussed by Bodine (in press).

Authigenic vs detrital origin

Several features of these assemblages suggest an authigenic origin for most, if not all clay minerals in these assemblages. First, although the two major clays, illite and chlorite, are common detrital constituents in a wide range of sedimentary rocks, their respective consistent compositions in the evaporite sections suggest an authigenic origin. The composition of the chlorite with its unusually high magnesium content (table 1) is more convincing than that of illite. Although the illites are well crystallized with high interlayer charge, and only modestly depleted in tetrahedral aluminum when compared with muscovite, their compositions do not differ markedly from some shale illites, e.g. the Silver Hill illite (table 1) reported by Srodoń and Eberl, in press. Furthermore, in most sedimentary rocks containing detritus from continental sources, chlorite is generally a minor phase. To be sure, accumulated detritus from a local source with these clay-mineral compositions could occur, e.g. weathered debris from a mafic rock terrain; however, the repetition of this association in time and space, and the large and varied nature of the source areas that supplied detritus to these basins would appear to rule out such a coincidence.

Second, the repeated, often frequent and abundant occurrence of minerals that only rarely occur in detrital accumulates strongly suggests an authigenic origin. The occurrence of talc at each of the three localities, the frequent occurrence of interstratified chlorite-saponite or corrensite at two localities, and the less frequent, but unequivocal occurrence of serpentine at two localities become exceedingly awkward to attribute to a detrital origin.

Finally, the chemical composition of clay-mineral assemblages from these evaporites (e.g. those from tables 2) are noticeably different from those observed for average shales (Pettijohn, 1975) and for clay-mineral assemblages from Gulf Coast Tertiary shales (Hower et al., 1976). The narrow range of bulk clay compositions from evaporites and their departure from the average detrital accumulates from continental sources (fig. 2) is convincing. The features on figure 2 reflect the dominance of dioctahedral clays (dioctahedral smectites, illite, interstratified illite-smectite, and kaolinite) with only minor quantities

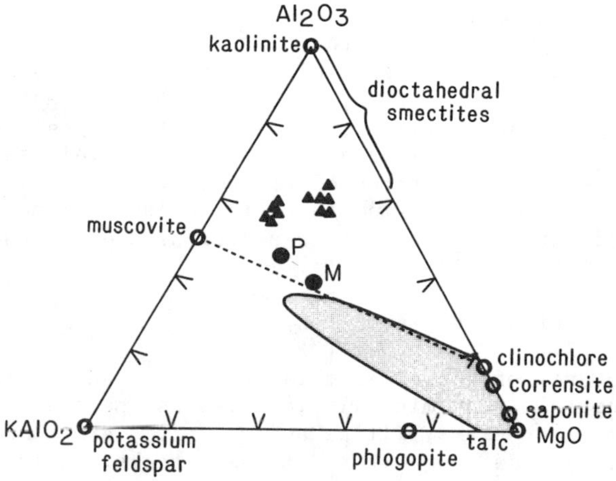

Fig. 2. Molar proportions (Al_2O_3-$KAlO_2$-MgO) in clay-mineral assemblages (68 samples) from Retsof, Paradox Basin, and Delaware Basin marine evaporite sections (stippled), contrasted with proportions in average Paleozoic, P, and Mesozoic, M, shales (Pettijohn, 1975), and clay-mineral assemblages from Gulf Coast Tertiary shales (Hower et al., 1976). Proportions in selected minerals and the clinochlore-muscovite join plotted for reference. Proximity of average shale proportions to the clinochlore-muscovite join reflects Mg and K enrichment by potassium feldspar and dolomite in the shales.

of trioctahedral clays in the average shale (the proximity of average shales to the clinochlore-muscovite join on fig. 2 is, in part, due to their feldspar and dolomite content) and Gulf Coast clay-mineral assemblages, as opposed to the abundance of trioctahedral clays and K-rich discrete illite in the marine evaporite assemblages.

Mechanisms of authigenesis

An authigenic origin of these clay-mineral assemblages requires destruction of the original detrital clays, as well as formation of the observed clays. Mechanisms are thought to include: (1) transformation of detrital dioctahedral clays (illite, smectite, interstratified illite-smectite, and kaolinite) into aggraded illite and the aluminous trioctahedral clays (discrete chlorite and interstratified chlorite-saponite, or more likely, their Mg-rich smectite precusors); (2) direct precipitation (neoformation) of talc and serpentine; (3) transformation of interstratified chlorite-smectite, or its Mg-enriched precursor into interstratified talc-saponite accompanying polyhalite replacing anhydrite; and (4) later isochemical or near-isochemical recrystallization of earlier authigenic clays to form the observed assemblages. Two of the mechanisms, the neoformation and distribution of talc and serpentine, and the formation of interstratified talc-saponite, have been previously reported (Bodine, in press) and will not be reviewed in this paper.

The remaining discussion will focus on transformation of the detrital clays during deposition and early diagenesis, and their subsequent isochemical recrystallization accompanying burial. In the former environment, the marine evaporite brine is the dominating factor; its composition is extreme and its volume with respect to the mass of argillaceous material is large. In the latter environment, because much of the pore fluid had been expelled during compaction, the volume of brine is relatively small, and mass transfer between the silicates and the pore fluid becomes negligible; major factors in this environment are increasing pressures and temperatures accompanying burial.

Transformation of detrital clays involves two reaction paths. One is formation of discrete illite; the second is transformation of Al-rich dioctahedral layers in detrital clays to Mg-enriched, Al-depleted layers in the transformed clays.

The discrete illite most likely formed from poorly crystallized detrital illites and interstratified illite-dioctahedral smectites. Early diagenetic additions of illite domains to mixed-layer clays have been suggested by Jones and Weir (1983) to occur in a modern alkaline lake environment. Similar illite growth and aggradation would not be unexpected in marine evaporite environments in which potassium concentration increases over that in seawater throughout all but the final stages of progressive evaporation (Braitsch, 1973). Theoretical stability relations (Garrels, 1984) show that at 25°C with increasing K/H activity ratios in the coexisting fluid, the solution leaves the interstratified illite-montmorillonite stability field, or as Garrels prefers, the illite-montmorillonite phase boundary, and enters the discrete illite field. His calculations further point out that low dissolved silica activity favors stability of discrete illite, even "muscovite"; the presence of any neoformed talc in the assemblage will tend to buffer silica activity to the talc-quartz equilibrium or lower. It appears reasonable that illite aggradation and "illitization" of interstratified illite-smectite occurs during near surface early diagenesis in the marine evaporite environment, and that the transformation probably continues during subsequent burial.

A second, and more extensive transformation is replacement of detrital dioctahedral smectites and kaolinite (detrital clays with no potassium) by the aluminous trioctahedral clays. An idealized, albeit kinetically difficult, example for such a transformation reaction is

$$\underset{\text{kaolinite}}{Al_2Si_2O_5(OH)_2} + 5Mg^{2+}_{aq} + H_4SiO_{4\,aq} + 5H_2O \rightarrow \underset{\text{clinochlore}}{(Mg_5Al)(Si_3Al)O_{10}(OH)_8} + 10H^{+}_{aq} \quad (1)$$

in which aluminum is conserved within solid phases. In this and ensuing reactions, the solid assemblage experiences no gain or loss of aluminum; concentrations of aluminum solutes, except at extreme pH's, are negligible and preclude significant mass transfer of aluminum between solids and the co-existing aqueous fluid. Reaction (1) transfers magnesium and aqueous silica from the brine into the solid, and, conversely, hydrogen ion from the solid into the brine. As long as the volume of brine with respect to the mass of argillaceous material is large, this appears reasonable. High magnesium concentration in the brine is the magnesium source; coexisting detrital quartz maintaining a continuing supply of dissolved silica is the aqueous silica source; and the buffering interaction among aqueous carbon dioxide species and a co-existing carbonate mineral consumes released hydrogen ion.

A remaining concern is whether or not thermodynamic and kinetic constraints permit the reaction to proceed spontaneously. Using 25°C free energy values (Robie, Hemingway and Fisher, 1981) and a calculated free-energy approximation (Tardy and Garrels, 1974) for clinochlore, reaction (1) is thermodynamically permissable. Kinetic constraints are less clear. Metastable kaolinite in marine sediments and sedimentary rocks, for example, survives for geologically significant time periods until the sediment is buried to considerable depth (Hower et al., 1976). If the transformation is to occur at or near surface conditions, excess potential from the high magnesium concentrations in the brine must overcome the kinetic barrier. Observed clay-mineral interactions with brines in modern alkaline lake environments (Khoury, Eberl, and Jones, 1982; Jones and Weir, 1983) suggest some kinetic constraints can be overcome in surficial environments.

In this manner the various dioctahedral reactants are believed to be transformed into an authigenic assemblage whose chemical compositions approximate those in the observed assemblages. At the same time, it appears doubtful that the mineralogy of these early authigenic assemblages was the same as in the observed assemblages. Initial transformation reactions, for example, may well have produced poorly crystallized Mg-enriched smectities such as have been observed by Jones and Weir (1983) in a modern alkaline lake environment. Their authigenic smectites are enriched in magnesium, exhibit an octahedral site occupancy midway between the dioctahedral and trioctahedral structures (2.4-2.5 octahedral cations per formula unit), and contain appreciable chlorite intergrades.

Observed variations in the mineralogy of the aluminous trioctahedral clays as a function of salinity, i.e. the occurrence of discrete chlorite and chlorite-rich interstratified chlorite-saponite in basal shale and carbonate intervals as opposed to interstratified chlorite-saponite or corrensite in anhydrite and halite intervals in the Paradox cycles (fig. 1), can be attributed to progressively increasing magnesium concentrations in the coexisting brines throughout each cycle. Just as reaction (1) is favored by increasing magnesium concentration, or, more properly, increasing $Mg^{2+}/(H^+)^2$ activity ratio, a second idealized reaction,

$$\underset{\text{clinochlore}}{2(Mg_5Al)(Si_3AL)O_{10}(OH)_8} + 3.67Mg^{2+}_{aq} + 5H_4SiO_{4\ aq} \rightarrow$$

$$\underset{\text{corrensite}}{1.67Mg_{0.2}(Mg_8Al)(Si_{6.6}Al_{1.4})O_{20}(OH)_{10}} + 7.33H^+_{aq} + 6H_2O \qquad (2)$$

continues the process and is similarly promoted by an increased $Mg^{2+}/(H^+)^2$ activity ratio. Thus, it should not be unexpected that a transition from discrete chlorite to corrensite appears to accompany progressive evaporation.

Parallel relations apply if the suggested Mg-rich smectite precursors, rather than clays in the chlorite-corrensite continuum, constitute the initial near-surface authigenic assemblage. The magnesium content, the excess octahedral cation site occupancy, and the extent of chlorite-intergrade growth are also functions of the $Mg^{2+}/(H^+)^2$ activity ratio. An idealized succession of smectites from a dioctahedral clay (montmorillonite) to a trioctahedral clay (stevensite) is shown in table 4, in which the increasing excess of dioctahedral site occupancy requires a concomitant increase of the Mg/Al atom ratio in the octahedral layer. The trioctahedral end member of the succession, stevensite, still retains octahedral-layer defects (omissions) from the trioctahedral structure in order to maintain an appropriate layer charge deficiency. The transformation reaction of a detrital montmorillonite to form a hypothetical authigenic smectite with an octahedral site occupancy of 2.5 cations per formula unit,

$$Mg_{0.2}(Al_{1.6}Mg_{0.4})Si_4O_{10}(OH)_2 + 5Mg^{2+}_{aq} + 6.67H_4SiO_{4\ aq} \rightarrow$$
montmorillonite

$$2.67Mg_{0.2}(Al_{0.6}Mg_{1.9})Si_4O_{10}(OH)_2 + 10H^+_{aq} + 6.68H_2O \quad (3)$$
Mg-enriched smectite

is promoted, like reactions (1) and (2) above, by an increased $Mg^{2+}/(H^+)^2$ activity ratio.

In either case, whether considering transformations directly to the observed chlorite-corrensite continuum or to their hypothetical Mg-rich smectite precursors, the extent of magnesium enrichment in the clay, i.e. the increase of the Mg/Al atom ratio, is a function of the $Mg^{2+}/(H^+)^2$ activity ratio in the coexisting ambient brine. Although the pH of marine evaporite brine decreases during the middle and latter stages of progressive evaporation (see Bodine, 1976, for example), as opposed to the behavior of alkaline playa brines, the $Mg^{2+}/(H^+)^2$

TABLE 4. Idealized formulae of detrital montmorillonite (1), two hypothetical Mg-enriched aluminous smectites (2 and 3), and the tri-octahedral end-member smectite, stevensite (4) with their cation occupancy per formula unit, and octahedral magnesium-aluminum atom ratio, $(Mg + Al)_{oct}$ and $(Mg/Al)_{oct}$, respectively

	Formula	$(Mg + Al)_{oct}$	$(Mg/Al)_{oct}$
(1)	$Mg_{0.2}(Al_{1.6}Mg_{0.4})Si_4O_{10}(OH)_2$*	2.0	0.25
(2)	$Mg_{0.2}(Al_{1.1}Mg_{1.15})Si_4O_{10}(OH)_2$	2.25	1.05
(3)	$Mg_{0.2}(Al_{0.6}Mg_{1.9})Si_4O_{10}(OH)_2$	2.5	3.17
(4)	$Mg_{0.2}(Mg_{2.8})Si_4O_{10}(OH)_2$#	2.8	∞

* Interlayer cation charge in smectites assigned a value of 0.4, rather than the more common value of 0.33, to facilitate later reaction balances; mechanisms proposed by Jones and Weir (1983) suggest addition of authigenic magnesium silicate to a detrital (dioctahedral) smectite results in a progressive decrease in layer charge as well as an increase in octahedral site occupancy.

Octahedral site deficiency in stevensite required to generate interlayer charge (see Eberl, Jones, and Khoury, 1982, for natural examples).

activity ratio continues to increase in the marine system at least through barren halite and into bitter salt precipitation.

In summary, it is likely that surface or near-surface transformation of detrital clays took place. Although not documented directly, it suggested that initial authigenic assemblages of poorly crystallized illite and Mg-enriched smectitic clays were the isochemical or near-isochemical precursors of the minerals in the present assemblages. It is most likely that most Mg-enrichment occurred during and shortly after deposition when brine volume with respect to the mass of argillaceous material was large and the brine behaved as an "infinite reservoir" of magnesium. Finally, the degree of Mg-enrichment in the precursors, whether they be hypothetical Mg-enriched smectites or poorly crystallized clays along the chlorite-corrensite continuum, appears to have been a function of progressive evaporation (increasing $Mg^{2+}/(H^+)^2$ activity ratio) in the host brine.

Clay authigenesis accompanying burial with gradually increasing temperatures over an extended period of geologic time completed formation of the observed clay-mineral assemblages. Because saline rocks characteristically have low porosity and brine volume is restricted, only negligible exchange between the solid assemblage and the host pore-fluid can occur without seriously altering fluid composition. Thus, idealized reactions accompanying burial are constructed with no fluid-solid exchange. Two reaction paths are involved: improved illite crystallinity; and isochemical or near-isochemical recrystallization of the Mg-enriched clays.

Little needs to be added regarding illite genesis other than restating that illite in these assemblages is both very well crystallized and, to some extent, compositionally distinctive. There is every likelihood that further development of these properties, particularly the former, continued during burial.

Several features among the trioctahedral clays deserve consideration. First, any authigenic Mg-enriched precursors that formed were transformed into the observed minerals. Secondly, clays from the Delaware Basin section, less deeply buried than the remaining two sections, are poorly crystallized and heterogeneous when compared with each cycle's progression from discrete chlorite to ordered 1:1 interstratified chlorite-saponite throughout the Paradox section. Finally, the absence of interstratified chlorite-smectite in the Retsof assemblages, the most deeply buried of the three localities, suggests instability of the trioctahedral smectite layers at greater depths.

A wide range of possible burial-metamorphic reactions can be constructed assuming that the Mg-enriched smectites, such as those described by Jones and Weir (1983), formed as initial authigenic replacements of the detrital assemblage. However, because the authigenic precursor compositions are unknown, only the most idealized reactions are warrented. These are shown in table 5 in two reactions (A and B) in which Mg-enriched smectites react to form idealized chlorite

TABLE 5. Idealized transformation reactions assuming no exchange with the coexisting pore fluid; reactions (A) and (B) transform Mg-enriched smectites into chlorite and corrensite, respectively; reactions (C), (D), and (E) transform corrensite, as suggested at Retsof, into chlorite with coexisting quartz (C), with coexisting quartz and talc, and retaining ideal clinochlore composition (D), and with coexisting talc (E)

(A) $$\underset{\text{Mg-enriched smectite}}{2.667Mg_{0.2}(Al_{0.75}Mg_{1.675})Si_4O_{10}(OH)_2} + 1.333H_2O \rightarrow \underset{\text{chlorite}}{(Mg_5Al)(Si_3Al)O_{10}(OH)_8} + \underset{\text{quartz}}{7.667SiO_2}$$

(B) $$\underset{\text{Mg-enriched smectite}}{3.933Mg_{0.2}(Al_{0.61}Mg_{1.885})Si_4O_{10}(OH)_2} + 1.067H_2O \rightarrow \underset{\text{corrensite}}{Mg_{0.2}(Mg_8Al)(Si_{6.6}Al_{1.4})O_{20}(OH)_{10}} + \underset{\text{quartz}}{9.133SiO_2}$$

- -

(C) $$\underset{\text{corrensite}}{Mg_{0.2}(Mg_8Al)(Si_{6.6}Al_{1.4})O_{20}(OH)_{10}} + 1.267H_2O \rightarrow$$
$$\underset{\text{chlorite}}{1.567(Mg_{5.234}Al_{0.766})(Si_{3.234}Al_{0.766})O_{10}(OH)_8} + \underset{\text{quartz}}{1.533SiO_2}$$

(D) $$\underset{\text{corrensite}}{Mg_{0.2}(Mg_8Al)(Si_{6.6}Al_{1.4})O_{20}(OH)_{10}} + 0.533H_2O \rightarrow$$
$$\underset{\text{chlorite}}{1.2(Mg_5Al)(Si_3Al)O_{10}(OH)_8} + \underset{\text{talc}}{0.733Mg_3Si_4(OH)_2} + \underset{\text{quartz}}{0.068SiO_2}$$

(E) $$\underset{\text{corrensite}}{Mg_{0.2}(Mg_8Al)(Si_{6.6}Al_{1.4})O_{20}(OH)_{10}} + 0.499H_2O \rightarrow$$
$$\underset{\text{chlorite}}{1.183(Mg_{4.986}Al_{1.014})(Si_{2.986}Al_{1.014})O_{10}(OH)_8} + \underset{\text{talc}}{0.767Mg_3Si_4O_{10}(OH)_2}$$

and corrensite compositions, respectively. The reactions involve no species other than water, silica, and the clay minerals. The precursor smectite compositions have octahedral site occupancies of 2.42 and 2.49 cations per formula unit respectively; these fall conveniently within the range observed by Jones and Weir (1983) with site occupancy of the smectite transforming to chlorite being the lesser of the two. These reactions are not being proposed as the specific reactions that occurred, but serve as one pair of examples for the hypothetical transformations. Too many unknown variables exist to write the reactions with any degree of confidence, e.g. the interlayer charge and composition of the precursor smectites, the composition, particularly the aluminum and ferrous iron content, of the resultant clays, and the extent of exchange, if any, with the host pore fluid.

Depths of burial at which these reactions take place are not known, but can be qualitatively estimated by mineralogic characteristics in two of the assemblages. Assemblages from the Paradox section are mature; corrensite is a common constituent, and discrete saponite and interstratified chlorite-saponite with greater than 50% saponite layers are absent. In the less deeply buried Delaware Basin assemblages, on the other hand, the trioctahedral clays are decidedly less well crystallized, saponite-rich clays, including discrete saponite, occur, and corrensite is rare. Whether the contrast between the two localities reflects different degrees of burial recrystallization or is related to greater postdepositional fluid mobility and salt recrystallization that occurred throughout the Salado Formation remains unresolved; however, it appears reasonable that depth of burial played a major role.

Another feature of the clay-mineral assemblages is the difference between clay assemblages from Paradox salts and those from Retsof. Interstratified chlorite-saponites are common in the former, but totally absent in the latter; well crystallized discrete chlorite occurs pervasively in the Retsof salt. If authigenic precursors formed during early diagenesis at both deposits and were subsequently transformed into interstratified chlorite-saponites in the more saline beds during progressive burial, then it must be concluded, that with deeper burial and higher temperatures as at Retsof, saponite layers became unstable and discrete chlorite formed.

The proposed transitions qualitatively parallel those observed in the burial metamorphism of dioctahedral clays. Hower et al. (1976) clearly document the progression in Gulf Coast shales. Randomly ordered interstratified illite-dioctahedral smectite in the upper part of the section (1850-3400 m depths) are ordered to the allevardite-type layer alternation in the 3400-3500 m-depth interval. Concomitantly, throughout the interval between 2150 and 3700 m (90°C borehole temperature at 3605 m), layer composition of the interstratified clay progressively increases from <40 to >80% illite in the finest size fraction. In a recent abstract, Mackenzie, Schoonmaker, and Chang (1983) describe analogous transformations for a trioctahedral clay succession in mafic shales from the Brazilian shelf. With increasing depth, saponite is transformed into interstratified chlorite-saponite with a progressive increase of chlorite-rich phases.

It is suggested that with increasing depth of burial, the trioctahedral clays in the evaporite assemblages follow a similar progression. Relatively shallow (low-temperature) crystallization of randomly ordered interstratified chlorite-trioctahedral smectites that are commonly preserved in the Delaware Basin section is followed by deeper (modestly elevated temperature) occurrences of partially and fully ordered (corrensite) interstratified trioctahedral clays in the Paradox salts, and culminates at greater depths (temperature $\sim 100^{o}C$) in Retsof salts with instability of the smectite layers

Recrystallization of corrensite into chlorite, like the earlier transformations, can occur through a wide variety of reactions. Three idealized reactions are shown, (C), (D), and (E) in table 5. The reactions are presented only as examples to demonstrate the feasibility of the transformation. The three reactions differ in that the recrystallization products are chlorite-quartz (C), chlorite-talc-quartz (D), and chlorite-talc (E); furthermore, reaction (D) was constructed to preserve an idealized clinochlore composition in the recrystallized product. Because talc appears to be a common neoformed constituent at all three localities and is particularly abundant in the Retsof assemblages, and because the residual pore fluids were most likely in equilibrium with talc or the talc-quartz pair (Bodine, in press), reactions (D) and (E) appear attractive. In addition, reactions (D) and (E) avoid producing a severely altered chlorite composition; the chlorite in reaction (A) is significantly Al-depleted, a penninite (Foster, 1962) rather than a clinochlore. Penninite compositions are rare and were not found for any Retsof chlorites (Bodine and Standaert, 1977). It should also be noted, that, if exchange did take place between the clays and co-existing pore fluid, the replacement of corrensite by chlorite likely followed the reverse of reaction (2).

Summary of conclusions

This paper stresses the importance of clay-brine interaction in the formation of the distinctive clay-mineral assemblages in marine evaporite rocks, and suggests the following sequence for their genesis.

1. Detrital argillaceous material deposited in evaporite basins characteristically consisted of the clays normally associated with continental weathering and fluvial transport, i.e. kaolinite, dioctahedral smectites, illite, and interstratified illite-dioctahedral smectite.

2. During early diagenesis in surface and near-surface environments, while brine composition reflected evaporative concentration and brine volume with respect to the mass of argillaceous material was large, detrital clays interacted with the brine's high magnesium and potassium concentrations and were transformed into an initial authigenic assemblage of aggraded illite and Mg-enriched precur-

sors. The precursors are suggested to have been Mg-enriched montmorillonites with octahedral site occupancies midway between dioctahedral and trioctahedral structures, and with their magnesium contents varying as a function of magnesium concentrations in the co-existing brines.

3. Isochemical or near-isochemical transformations accompanying burial to moderate depths resulted in continued aggradation of illite and formation of poorly crystallized trioctahedral smectites (saponites), poorly ordered interstratified chlorite-saponites, and discrete chlorite with chorite-rich clays dominating penesaline intervals and more smectitic clays dominating hypersaline intervals.

4. Deeper burial and elevated temperatures resulted in increased crystallinity of the clays and, in evaporite intervals with smectitic clays, their transformation to more regularly ordered species with 1:1 regularly ordered corrensite frequently forming.

5. With further burial and increased temperature (~100°C), the stability limit of the trioctahedral smectite layers was exceeded, and corrensite and other interstratified clays were transformed into chlorite with excess magnesium and silica forming co-existing talc and quartz.

REFERENCES

Bailey, S. W., 1982, "Nomenclature for Regular Interstratifications," American Mineralogist, Vol. 67, nos. 3 and 4, Mar.-Apr., pp. 394-398.

Bodine, M. W., Jr., 1976, "Magnesium Hydroxychloride: A Possible pH Buffer in Marine Evaporite Brines?" Geology, Vol. 4, No. 2, Feb., pp. 76-80.

Bodine, M. W. Jr., 1978, "Clay-Mineral Assemblages from Drill Core of Ochoan Evaporites, Eddy County," Geology and Mineral Deposits of Ochoan Rocks in Delaware Basin and Adjacent Areas, Austin, G. S., ed., Circular 159, New Mexico Bureau of Mines and Mineral Resources, pp. 21-31.

Bodine, M. W., Jr., in press, "Trioctahedral Clay Mineral Assemblages in Paleozoic Marine Evaporite Rocks," Sixth International Symposium on Salt.

Bodine, M. W., Jr. and Fernalld, T. H., 1973, "EDTA Dissolution of Gypsum, Anhydrite, and Ca-Mg Carbonates," Journal of Sedimentary Petrology, Vol. 43, no. 4, Dec., pp. 1152-1156.

Bodine, M. W., Jr. and Rueger, B. F., 1984, "Progress Report on Clay-Mineral Assemblages in the Gibson Dome No. 1 Drill Core, Paradox Basin, Utah," Open-File Report 84-165, U.S. Geological Survey.

Bodine, M. W., Jr. and Standaert, R. R., 1977, "Chlorite and Illite Compositions from Upper Silurian Rock Salts, Retsof, New York," Clays and Clay Minerals, Vol. 25, No. 2, May, pp. 57-71.

Braitsch, Otto, 1973, Salt Deposits; Their Origin and Composition, Springer-Verlag, New York.

Brown, George and Brindley, G. W., 1980, "X-ray Diffraction Procedures for Clay Mineral Identification," Crystal Structures of Clay Minerals and their Identification, Brindley, G. W. and Brown, George, eds., Mineralogical Society, London, pp. 305-359.

Case, W. S., Jr., 1978, "Leach/Crystallization-Process for Potash Production," Geology and Mineral Deposits of Ochoan Rocks in Delaware Basin and Adjacent Areas, Austin, G. S., ed., Circular 159, New Mexico Bureau of Mines and Mineral Resources, pp. 69-70.

Cheeseman, R. J., 1978, "Geology and Oil/Potash Resources of Delaware Basin, Eddy and Lea Counties, New Mexico," Geology and Mineral Deposits of Ochoan Rocks in Delaware Basin and Adjacent Areas, Austin, G. S., ed., Circular 159, New Mexico Bureau of Mines and Mineral Resources, pp. 7-14.

Dean, W. E., Davies, G. R., and Anderson, R. Y., 1975, "Sedimentological Significance of Nodular and Laminated Anhydrite," Geology, Vol. 3, No. 7, July, pp. 367-372.

Dellwig, L. F. and Evans, R., 1969, "Depositional Processes in Salina Salt of Michigan, Ohio, and New York," American Association of Petroleum Geologists Bulletin, Vol. 53, No. 4, April, pp. 949-956.

Eberl, D. D., Jones, B. F., and Khoury, H. N., 1982, "Mixed-Layer Kerolite-Stevensite from the Amargosa Desert, Nevada," Clays and Clay Minerals, Vol. 30, No. 5, pp. 321-326.

Foster, M. D., 1962, "Interpretation of the Composition and a Classification of the Chlorites," Professional Paper 414-Y, U.S. Geological Survey.

Füchtbauer, Hans and Goldschmidt, Hertha, 1959, "Die Tonminerale der Zechsteinformation," Beiträge zur Mineralogie und Petrographie, Vol. 6, pp. 320-345.

Garrels, R. M., 1984, "Montmorillonite/Illite Stability Diagrams," Clays and Clay Minerals, Vol. 32, No. 3, June, pp. 161-166.

Harris, A. G., Harris, L. D., and Epstein, J. B., 1978, "Oil and Gas Data from Paleozoic Rocks in the Appalachian Basin: Maps for Assessing Hydrocarbon Potential and Thermal Maturity (Conodont Color Alteration Isograds and Overburden Isopachs)," Miscellaneous Map Series, U.S. Geological Survey, Map I-917-E.

Hite, R. J., 1960, "Stratigraphy of the Saline Facies of the Paradox Member of the Hermosa Formation of Southeastern Utah and Southwestern Colorado," Geology of the Paradox Basin Fold and Fault Belt, 3rd Field Conference, Durango, Colorado, Four Corners Geological Society, pp. 86-89.

Hite, R. J., 1983, "Preliminary Mineralogical and Geochemical Data from the D.O.E. Gibson Dome Corehole No. 1, San Juan County, Utah," Open-File Report 83-780, U.S. Geological Survey.

Hite, R. J. and Buckner, D. H., 1981, "Stratigraphic Correlations, Facies Concepts, and Cyclicity in Pennsylvanian Rocks of the Paradox Basin," Geology of the Paradox Basin, Wiegand, D. L., ed., Field Conference, Rocky Mountain Association of Geologists, pp. 147-160.

Hower, John et al., 1976, "Mechanism of Burial Metamorphism of Argillaceous Sediment: Mineralogical and Chemical Evidence," Geological Society of America Bulletin, Vol. 87, No. 5, May, pp. 725-737.

Jones, B. F. and Weir, A. H., 1983, "Clay Minerals of Lake Abert, An Alkaline, Saline Lake," Clays and Clay Minerals, Vol. 31, No. 3, June, pp. 161-172.

Jones, C. L., 1954, "Potash Resources in part of the Los Medanos Area of Eddy and Lea Counties, New Mexico," Open-File Report 75-407, U.S. Geological Survey.

Khoury, H. N., Eberl, D. D., and Jones, B. F., 1982, "Origin of Magnesian Clays from the Amargosa Desert, Nevada," Clays and Clay Minerals, Vol. 30, No. 5, Oct., pp. 327-336.

Mackenzie, F. T., Schoonmaker, Jane, and Chang, H. 1983, "Diagenesis and Mass Transfer in Sandstone-Shale Sequences," (Abstract), American Association of Petroleum Geologists Bulletin, Vol. 67, No. 3, Mar., p. 508.

Mayrhofer, Heimo, 1955, "Beitrage zur Kenntnis des Alpinen Salzgebirges," Zeitschrift Deutsche Geologische Gesellschaft (Hannover), Vol. 105, pp. 752-775.

Niemann, Hans, 1960, "Untersuchungen am Grauen Salzton der Grube 'Königshall-Hindenburg,' Reyerhausen bei Göttingen," Beiträge zur Mineralogie und Petrographie, Vol. 6, pp. 320-345.

Perry, E. A., Jr, 1971, "Silicate-Sea Water Equilibria in the Ocean System: A Discussion," Deep-Sea Research, Vol. 18, pp. 921-924.

Pettijohn, F. J., 1975, Sedimentary Rocks, 3rd edition, Harper and Row.

Powers, D. W. et al., 1978, "Geological Characterization Report, Waste Isolation Pilot Plant (WIPP) Site, Southeastern New Mexico," SAND

78-1596, Sandia Laboratories, Albuquerque, New Mexico, 2 Vols.

Raup, O. B., 1966, "Bromine Distribution in some Halite Rocks of the Paradox Member, Hermosa Formation, Utah," Second Symposium on Salt, Rau, J. L., ed., Northern Ohio Geological Society, Vol. 1, pp. 236-247.

Raup, O. B., Hite, R. J., and Groves, H. L., Jr., 1970, "Bromine Distribution and Paleosalinities from Well Cuttings, Paradox Basin, Utah and Colorado," Third Symposium on Salt, Rau, J. L. and Dellwig, L. F., eds., Northern Ohio Geological Society, Vol. 1, pp. 40-47.

Reynolds, R. C., 1980, "Interstratified Clay Minerals," Crystal Structures of Clay Minerals and their Identification, Brindley, G. W. and Brown, George, eds., Mineralogical Society, London, pp. 249-303.

Rickard, L. V., 1969, "Stratigraphy of the Upper Silurian Salina Group, New York, Pennsylvania, Ohio, Ontario," Map and Chart Series 12, New York State Museum and Science Service.

Robie, R. A., Hemingway, B. S., and Fisher, J. R., 1978, "Thermodynamic Properties of Minerals and Related Substances at 298.15 K and 1 Bar (105 Pascals) Pressure and at Higher Temperatures," Bulletin 1452, U.S. Geological Survey.

Saberian, A., 1980, "Accomplishments of SMRI-Sponsored Salt Dissolution Research since the Fourth Symposium on Salt," Fifth Symposium on Salt, Coogan, A. H. and Hauber, Lukas, eds., Northern Ohio Geological Society, pp. 147-159.

Srodoń, Jan and Eberl, D. D., in press, "Illite," Mica, Short Course Notes, Mineralogical Socoety of America, Chapt. 12.

Stewart, F. H., 1963, "Marine Evaporites," Data of Geochemistry, Fleischer, Michael, ed., Professional Paper 440-Y, U.S. Geological Survey.

Taggert, J. E., Jr., Lichte, F. E., and Wahlberg, J. S., 1981, "Methods of Analysis of Samples Using X-Ray Fluorescence and Induction-Coupled Plasma Spectroscopy," The 1980 Eruptions of Mount St. Helens, Washington, Lippman, P. W. and Mullineaux, D. R., eds., Professional Paper 1250, U.S. Geological Survey, pp. 683-687.

Tardy, Yves and Garrels, R. M., 1974, "A Method of Estimating the Gibbs Energies and Formation of Layer Silicates," Geochimica et Cosmochimica Acta, Vol. 38, No. 7, July, pp. 1101-1116.

Weaver, C. E. and Pollard, L. D., 1973, The Chemistry of Clay Minerals, Elsevier, New York.

Chapter 10

MINERALOGY OF SOME EGYPTIAN BENTONITIC-CLAY DEPOSITS

by Prof. Dr. Abdel Kadir Attia,* Prof. Dr. Haydn H. Murray,** Prof. Dr. Adel Abdel Azin* and Dr. Sabah N. Boulis*

ABSTRACT

An extensive reconnaissance survey of bentonitic clays in Egypt was carried out from 1982 to 1984. X-ray diffraction, differential thermal analysis, electron microscopy, and physical property evaluations were used to identify the mineralogy and assess their industrial potential.

Samples were collected from a number of localities, namely: Qasr El-Sagha (Fayum), Cairo-Alexandria Desert Road (Western Desert), and Qattamia area (Eastern Desert).

This paper reviews the geological occurrence, the mineralogical and chemical characteristics, and the potential utilization of these deposits.

INTRODUCTION

Bentonitic clay deposits occur at a number of localities in Egypt. The most important includes Qasr El-Sagha, Qattamia, and an area along the Cairo-Alexandria Desert road (Fig. 1).

The Qasr El-Sagha deposits lie in the Fayum depression about 70 km south-southwest of Cairo. The mineralogy of these deposits was first studied by Basta, et al. (1971), and later by Hegab (1974). The section in which the bentonite occurs reaches about 200 m in thickness and consists mainly of sandstones, limestones and clays of Upper Eocene Age.

---

* Central Metallurgical Research and Development Institute, National Research Center, Dokki, Cairo, Egypt

** Department of Geology, Indiana University, Bloomington, Indiana 47405, U.S.A.

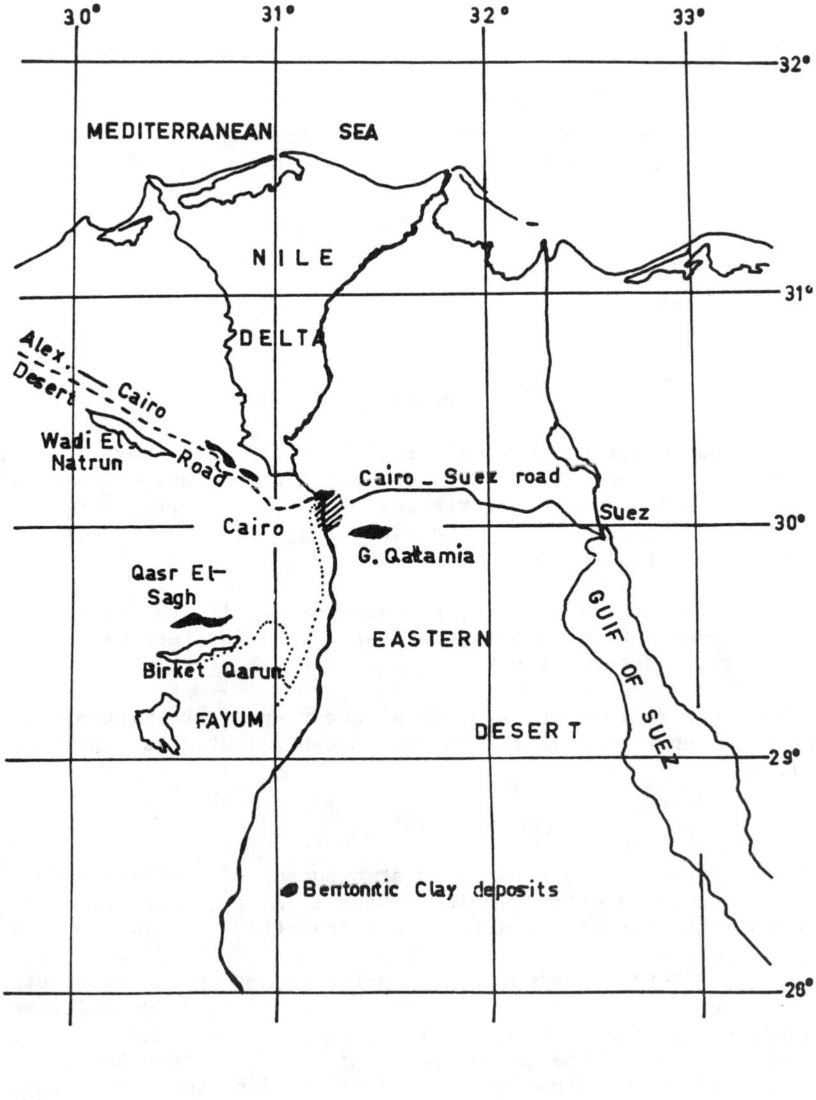

Fig. 1. Location Map of Egyptian Clay Deposits.

The Qattamia clay deposits are in the Maady area about 35 km southeast of Cairo (Fig. 1) These bentonites are intercalated with limestones of upper Eocene age and range in thickness between 1 to 4 meters (Figs. 2 and 3).

The clay deposits which lie along both sides of the Cairo-Alexandria desert road were recently discovered by the writers in the area along the road between kilometers 38 to kilometer 60. The excavations and fills for the new road exposed the bentonite at several locations. A number of pits were excavated in order to estimate the thickness of the clay beds and calculate the reserves. The thickness of the bentonite bed is estimated to average about 2 meters (Figs. 4 and 5), while the overburden, which consists mainly of loose sand, is about 1/2 to 1 meter in thickness.

Fig. 2. Test Pit, showing the bentonitic clay bed (about 3 m thick), Qattamia area.

Fig. 3. Outcrop of bentonitic clay (more than 3 m thick), Qattamia area.

Fig. 4. View showing the outcrop of the upper and lower bentonitic clay beds, along the face of a trench. Cairo-Alexandria Desert Road, km 48.

Fig. 5. Lower bentonitic clay bed (about 2 m thick), Cairo-Alexandria Desert Road.

SAMPLES COLLECTED

A number of representative samples were collected from the three deposits. The following table shows the exact locations of the samples and their lithology.

TABLE I. Bentonite Clay Sample Locations and Description

Sample	Locality	Description
	Area I	
F1	Lower Bed	40-ton technological sample, representing the lower bentonitic bed.
F2	Lower Bed	Greenish-gray to dark gray claystone, abundant gypsum bands, 1 1/2 m thick.
F3	Upper Bed	Grayish-brown, green in parts, claystone with some iron oxides staining in parts, gypsum bands are abundant, 1 1/2 m thick.

Table I, Cont'd.

Sample	Locality	Description
	Area II	
F4	Locality 1 Lower Bed	Greenish-gray to dark gray claystone, highly intercalated with gypsum, 1 1/4 m thick.
F5	Upper Bed	Greenish-gray claystone, abundant iron staining and gypsum intercalation.
F6	Locality 2	Greenish-gray to dark gray claystone, with gypsum intercalations.
	km 38 Cairo-Alex. desert road	
C1	Upper Bed	Greenish-gray claystone bed about 2 m thick.
C2	Lower Bed	Greenish-gray claystone bed, undetermined thickness.
C3	km 48 Upper Bed	Gray claystone with some organic material about 50 cm thick.
C4	Lower Bed	Gray claystone bed with some organic material intercalations, about 2 m in thickness.
C5	km 60	Greenish-gray claystone, completely free from gypsum and organic material, of unidentified thickness.
Q1	km 2.7	Gray claystone bed, 20 cm thick.
Q2	km 2.7	Greyish-brown claystone bed, 30 cm thick.
Q3	km 14.2	Gray, flaky, claystone bed, 1 m thick.
Q4	km 18.2	Grayish-green claystone bed, some organic material, about 3 m thick.
Q5	km 18.9	Gray claystone bed, 1 m thick.
Q6	km 21	Varigated color claystone bed, 1 m thick.

Table I, Cont'd.

Sample	Locality	Description
Q7 & Q7a	km 25 W Osman Quarry	Gray claystone bed, 1 m thick.
Q8	Osman Quarry Level III	Dark gray clay bed, 60 cm thick.
Q9	km 49 (the connection)* Section B	Gray color claystone bed, about 3 m thick.
Q10	km 50 Subtrack near Section B	Varicolored claystone bed, about 4 m thick.

F = Fayum C = Cairo-Alex. Desert Road Q = Qattamia

* The connection road between E. Maady-Qattamia and Cairo-Suez road.

EXPERIMENTAL PROCEDURES

Particle size analyses of the samples were carried out using the pipette method outlined by Kilmer and Alexander (1949), with some modification relative to the type of dispersing agent. The samples were first treated with dilute HCl (2%) and hydrogen peroxide in order to remove the secondary carbonates, the soluble salts and the organic matter. Each fraction was dried at about 100°C and weighed. The resulting weights were expressed as percentages of the total sample weight. Histograms, cumulative and triangular plots of sand, silt and clay percentages were constructed (Tables II to IV, and Figures 6 and 7), using an Apple II micro-computer.

Clay minerals collected from the different localities were examined by X-ray diffraction, differential thermal analysis, thermogravimetry, scanning electron microscopy, and were chemically analyzed. The X-rav diffraction patterns were run on a Philips powder diffractometer with monochromatized Cu Kα radiation, with a scanning speed of 1° 2θ per minute and operated at 40 Kv and 25 mA.

The sample was split into two aliquots, one for bulk mineral analysis and the other for clay analysis. The aliquot for bulk analysis was finely ground, mounted randomly on an aluminum holder, and analyzed by X-rav diffraction. The diffractograms were used to provide background information on the bulk mineralogy of the sample, and particularly to estimate relative abundance of non-clay minerals in the samples. The other aliquot from the same sample was disaggregated and dispersed in deionized water. The clay fractions (<2 um) were deposited as oriented aggregates on glass slides. Three diffractograms were obtained for each sample; namely, untreated, glycolated, and heated at 550°C for two hours. Scanning electron micrographs were obtained for the same clay fraction used in the X-ray diffraction studies.

DTA, TGA and DTG were recorded simultaneously on equipment manufactured by Netsch. The sample was ground to -200 mesh (74 µm) and alumina was used as a reference inert material. All runs were made at a heating rate of 10°C/min.

Complete chemical analyses and determination of exchangeable cations, e.g., Mg, Ca, Na and K, were done in the Geology Department at Indiana University, using plasma spectroscopy.

RESULTS AND DISCUSSION

Size Analysis

The histograms and cumulative curves (Figs. 6, 7, Tables II and III) indicate that these samples are fine grained. The median (Md) values vary between 1.7 µ and 2.6 µ in Qasr El-Sagha clays; between 1.9 µ and 26 µ in Cairo-Alexandria Desert road clays; and between 1.5 µ and 62 µ in Qattamia clays. However, considerable amounts of fine sand were detected in some of the samples, e.g., samples F5, C3, C4, C5, Q3 and Q10.

The clay fraction ranges between 47.5% and 81% in Qasr El-Sagh clays, between 22% and 58% in Cairo-Alexandria Desert road clays, and between 52% and 86% in Qattamia clays (Table II).

Table III and Figure 6 show the cumulative data obtained with the aid of an Apple II micro-computer. Table IV shows the sand-silt-clay calculated at 100%, the mean, median, degree of sorting, skewness, kurtosis, and the standard deviation for each sample. Figure 7 shows the sand-silt-clay sizes calculated to 100% and presented in a triangular diagram. From this figure it can be seen that three of the six Fayum clays and four of the Qattamia clays lie in the claystone category. Two of the Fayum clays, four of the Qattamia clays and two of the Cairo-Alexandria Desert road clays are classified as silty claystones. One sample from the Cairo-Alexandria Desert road suite is classified as sandy claystone. One of the Fayum clays and two of the Cairo-Alexandria Desert road samples are classified as clayey mudstones. Two of the Cairo-Alexandria Desert road clays are classified as clayey siltstones.

TABLE II. Grain-Size Data of the Bentonitic Clays

Sample No.	Fine Sand 125 μ	V.F. Sand 62 μ	V.C. Silt 31 μ	C. Silt 16 μ	M. Silt 8 μ	F. Silt 4 μ	V.F. Silt 2 μ	Clay <2 μ
F1	----	3.00	3.70	17.30	2.00	2.30	4.20	67.50
F2	----	0.50	1.00	3.00	6.00	6.00	8.00	75.50
F3	----	1.00	2.50	5.50	4.50	5.00	6.50	75.00
F4	----	1.20	5.80	8.00	2.00	0.50	1.50	81.00
F5	3.00	8.00	10.00	20.00	4.20	2.50	4.80	47.50
F6	----	5.00	3.20	9.80	4.70	3.30	4.00	70.00
C1	----	3.50	4.00	14.50	8.00	6.00	6.00	58.00
C2	1.00	1.50	12.00	13.00	8.50	6.00	9.00	49.00
C3	5.00	13.00	21.00	25.00	6.00	3.00	5.00	22.00
C4	2.00	12.00	13.00	25.00	9.00	6.00	5.00	28.00
C5	8.00	18.00	16.50	8.50	4.00	2.50	3.50	39.00
C6	----	1.50	11.00	20.00	5.00	3.50	4.00	55.00
Q1	----	-----	3.50	8.50	18.00	3.00	2.00	65.00
Q2	----	-----	3.50	2.80	19.50	12.00	5.20	57.00
Q3	3.50	8.50	8.00	4.00	3.00	1.00	2.00	70.00
Q4	----	-----	4.50	2.00	4.50	2.00	0.50	86.50
Q5	----	1.00	4.50	2.00	3.00	3.50	3.50	82.50
Q7	----	-----	2.00	1.50	7.00	5.00	4.50	80.00
Q7a	----	-----	2.00	5.50	10.50	4.50	2.50	75.00
Q8	----	-----	2.60	0.90	6.50	4.50	1.00	84.50
Q9	----	-----	-----	12.50	5.50	22.00	8.00	52.00
Q10	5.00	7.50	7.50	1.00	1.50	1.00	2.00	74.00

TABLE III. Cumulative Grain-Size Data of the Bentonitic Clays

Ø Sample	+3	+4	+5	+6	+7	+8	+9
F1	0.00	3.00	6.70	24.00	26.00	28.30	32.50
F2	0.00	0.50	1.50	4.50	10.50	16.50	24.50
F3	0.00	1.00	3.50	9.00	13.50	18.50	25.00
F4	0.00	1.20	7.00	15.00	17.00	17.50	19.00
F5	3.00	11.00	21.00	41.00	45.20	47.70	52.50
F6	0.00	5.00	8.20	18.00	22.70	26.00	30.00
C1	0.00	3.50	7.50	22.00	30.00	36.00	42.00
C2	1.00	2.50	14.50	27.50	36.00	42.00	51.00
C3	5.00	18.00	29.00	64.00	70.00	73.00	78.00
C4	2.00	14.00	27.00	52.00	61.00	67.00	72.00
C5	8.00	26.00	42.50	51.00	55.00	57.50	61.00
C6	0.00	1.50	12.50	32.50	37.50	41.00	45.00
Q1	0.00	0.00	3.50	12.00	30.00	33.00	35.00
Q2	0.00	0.00	3.50	6.30	25.80	37.80	43.00
Q3	3.50	12.00	20.00	24.00	27.00	28.00	30.00
Q4	0.00	0.00	4.50	6.50	11.00	13.00	13.50
Q5	0.00	1.00	5.50	7.50	10.50	14.00	17.50
Q7	0.00	0.00	2.00	3.50	10.50	15.50	20.00
Q7a	0.00	0.00	2.00	7.50	18.00	22.50	25.00
Q8	0.00	0.00	2.60	3.50	10.00	14.50	15.50
Q9	0.00	0.00	0.00	12.50	18.00	40.00	48.00
Q10	5.00	12.50	20.00	21.00	22.50	23.50	25.50

TABLE IV. Grain Size Data, Sorting (SO), Skewness (Sk), Kustosis (K), of the Bentonitic Clays

Sample No.	Sand	Silt	Clay	Mean	Median	So	Sk	K	St. Deviation
F1	3.00	29.50	67.50	10.10	1.80	21	3.03	13.09	18.47
F2	0.50	24.00	75.50	4.09	1.60	15	8.32	76.25	8.46
F3	1.00	24.00	75.00	5.55	1.70	20	4.71	28.36	12.31
F4	1.20	17.80	81.00	7.23	1.70	50	3.33	15.67	15.07
F5	11.00	41.50	47.50	23.88	2.60	32	2.59	10.20	38.88
F6	5.00	25.00	70.00	10.40	1.70	7.5	3.00	11.45	21.48
C1	3.50	38.5	58.00	10.81	1.90	24	2.98	12.41	19.26
C2	2.50	48.5	49.00	14.28	2.00	30	3.87	24.16	25.20
C3	18.00	60.0	22.00	38.55	26.00	62	1.89	6.52	44.84
C4	14.00	58.0	28.00	28.90	18.00	60	2.11	8.18	36.77
C5	26.00	35.0	39.00	42.80	16.00	90	1.44	4.25	54.58
C6	1.5	43.0	55.0	12.97	1.90	25	1.96	7.58	17.92
Q1	0.00	35.00	65.00	7.00	1.70	15.45	2.37	8.82	10.11
Q2	0.00	43.00	57.00	6.32	1.85	11.45	3.05	13.08	8.21
Q3	12.00	18.00	70.00	20.70	62.00	8.50	2.62	9.71	41.79
Q4	0.00	13.50	86.50	4.53	1.85	12.50	3.62	15.13	9.84
Q5	1.00	16.50	82.50	5.42	1.60	19.50	4.35	24.00	13.23
Q7	0.00	20.00	80.00	3.75	1.50	12.50	4.62	26.16	7.17
Q7a	0.00	25.00	75.00	4.95	1.60	15.50	3.21	14.48	8.23
Q8	0.00	15.50	84.50	3.76	1.70	20.00	4.57	24.58	7.74
Q9	0.00	48.00	52.00	5.93	1.80	15.50	1.73	4.58	7.17
Q10	12.50	13.00	74.50	21.52	1.88	87.00	2.57	8.90	46.06

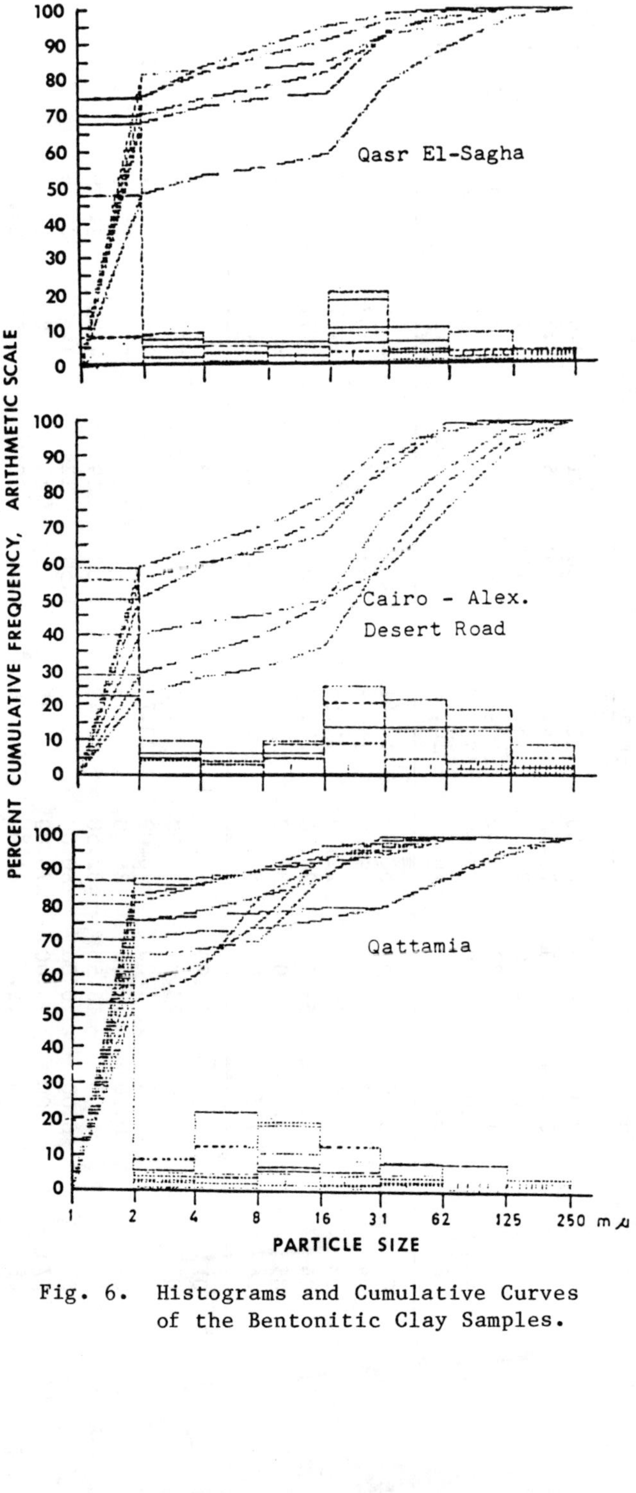

Fig. 6. Histograms and Cumulative Curves of the Bentonitic Clay Samples.

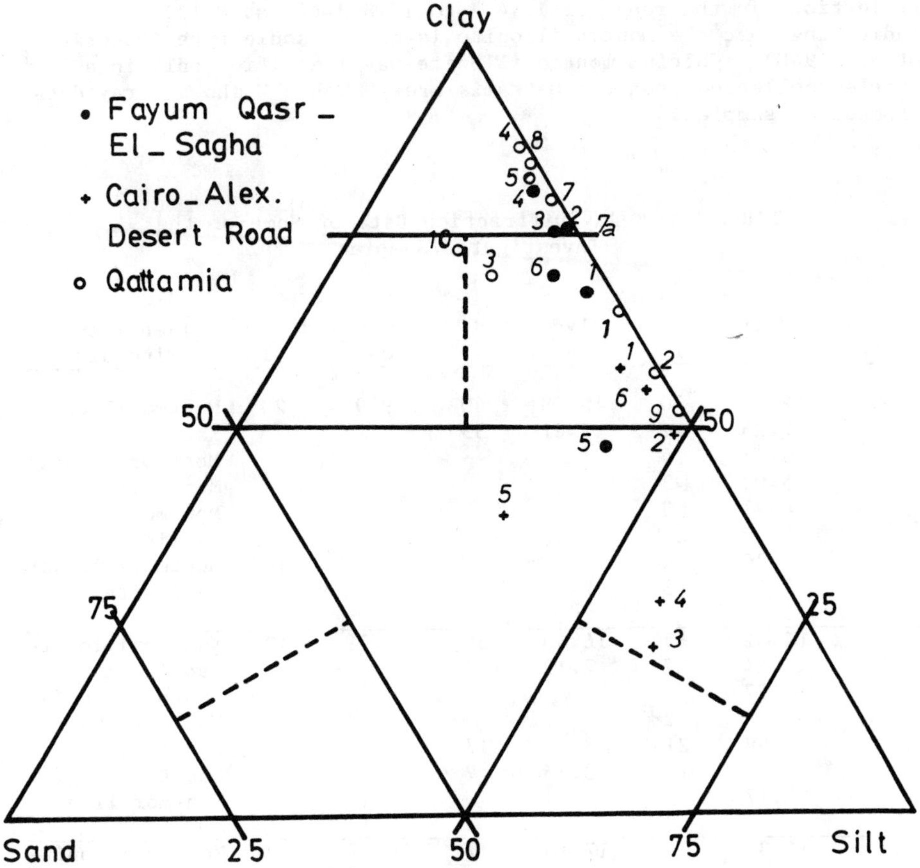

Fig. 7. Sand-Silt-Clay Percentages of the Bentonitic Clay Samples.

Mineralogical Composition

X-ray analysis of the samples revealed that montmorillonite is the most ubiquitous mineral. It was identified by an integral series of basal reflections near 17 Å after treatment with ethylene glycol. Air dried aggregates of 17 of 20 samples gave a series of reflections in the range of 12.4 Å to 12.8 Å (Figs. 8-12), thus indicating that the montmorillonite is of the sodic type (Hendricks, et al. 1940) Calcium montmorillonite was identified only in a few samples collected from the Qattamia area. Table V shows X-ray data from seven samples.

TABLE V. X-Ray Diffraction Data of Some of the Investigated Samples

Sample No.	Untreated dA	Untreated I	Glycolated dA	Glycolated I	Heated dA	Heated I	Identified Minerals
F2	12.82	70	16.28	40	9.9	23	Montmorillonite
	7.25	42	7.21	33			Kaolinite
	6.31	4					Montmorillonite
	5.93	3					Montmorillonite
	4.27	3					Montmorillonite
	3.58	20	3.57	19			Kaolinite
	3.26	4			3.21	10	Montmorillonite
	3.15	3					Kaolinite
F3	12.52	65	16.60	56	9.79	19	Montmorillonite
	7.24	42	7.19	30			Kaolinite
	6.27	3					Montmorillonite
	4.27	2					Montmor.,Quartz
	3.58	21	3.58	17			Kaolinite
	3.41	5	3.35	7	3.41	5	Quartz
	3.17	3			3.19		Montmorillonite
C2	12.37	16	17.10	17	10.07	10	Montmorillonite
	10.00	4	10.07	4			Illite
	7.14	12	7.24	11			Kaolinite
	3.57	9	3.56	7			Kaolinite
	3.34	3	3.34	2			Quartz
C3	12.83	20	17.10	26	10.08	10	Montmorillonite
	7.24	10	7.14	5			Kaolinite
	4.13	3					Montmorillonite
	3.57	7					Kaolinite
	3.34	5					Quartz

Table V, Cont'd.

Sample No.	Untreated dA	I	Glycolated dA	I	Heated dA	I	Identified Minerals
C4	12.11	46	16.67	22	9.90	10	Montmorillonite
	9.83	6	10.05	4			Illite
	7.14	18	7.14	10			Kaolinite
	3.56	11	3.57	7			Kaolinite
	3.33	7	3.33	3	3.33	5	Illite,Quartz
Q2	12.45	62	16.99	53	9.88	51	Montmorillonite
	7.13	75	7.13	42			Kaolinite
			5.57	3			Montmorillonite
	4.98	4			4.93	8	Montmorillonite
	4.25	4	4.23	3			Montmorillonite
			3.79	3			
	3.56	52	3.56	35			Kaolinite
					3.50	4	
	3.33	12	3.33	11	3.33	18	Illite
					3.22	23	
	2.38	4	2.38	3			Montmorillonite
Q7	14.85	135	16.99	88	10.16	90	Montmor.,Illite
			9.51	4			Illite
			8.84	7			Montmor.,Illite
	7.08	60	7.08	39			Kaolinite
					4.87	12	Montmor.,Illite
					4.82	12	Montmorillonite
	4.25	4	4.25	4	4.27	6	Montmorillonite
	3.56	40	3.56	32			Kaolinite
	3.34	11	3.34	20	3.35	25	Illite,Quartz
					3.22	39	

Kaolinite was identified by the (001) and (002) basal reflections near 7.2 Å and 3.55 Å. Heating the oriented specimens to 550°C completely destroyed these two reflections. The kaolinite content of the samples ranges from 13.7% to 53%.

Traces of montmorillonite-illite mixed layer and illite minerals were identified in the Qattamia and Cairo-Alexandria Desert road clay deposits. Quartz and feldspars are the main impurities in the Cairo-Alexandria Desert road samples. Gypsum is the main impurity in the Qasr El-Sagha and Qattamia deposits.

Electron microscopic examination of the samples showed that montmorillonite particles occur in a variety of forms, i.e., compact lamellar aggregates (Fig. 13), globular (Otay-type) aggregates (Fig. 14), and moss-like aggregates (Fig. 15). Kaolinite was observed in nearly all samples in the form of hexagonal platelets.

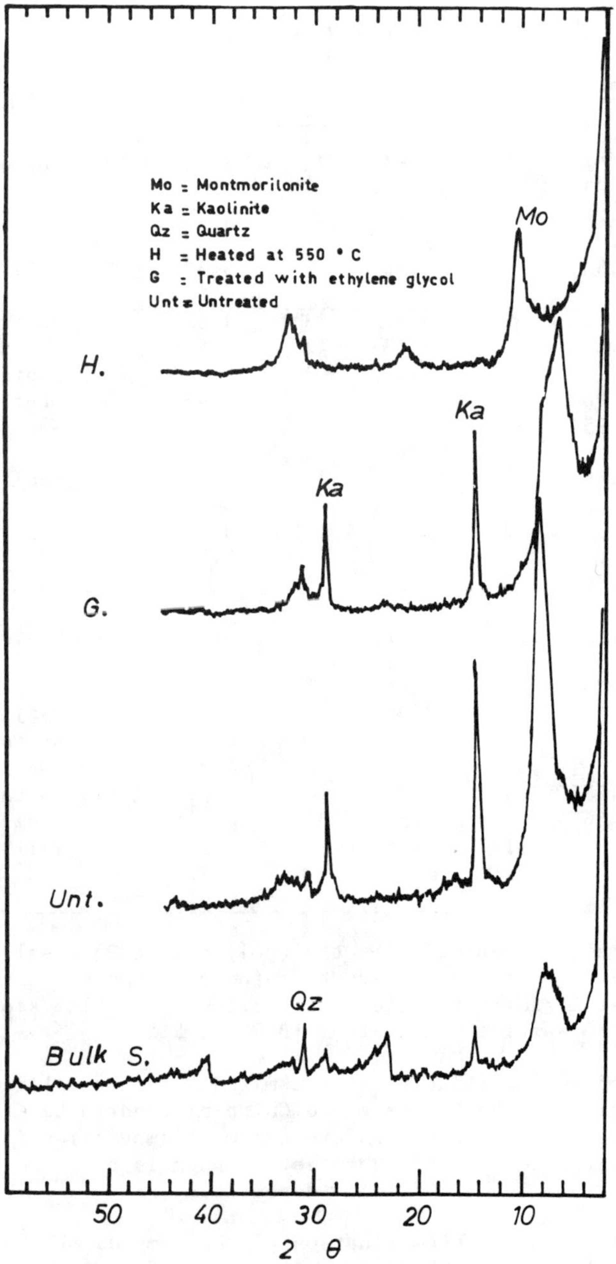

Fig. 8. X-Ray Diffractograms of the Lower Bentonitic Clay Bed, Qasr El-Sagh, Fayum.

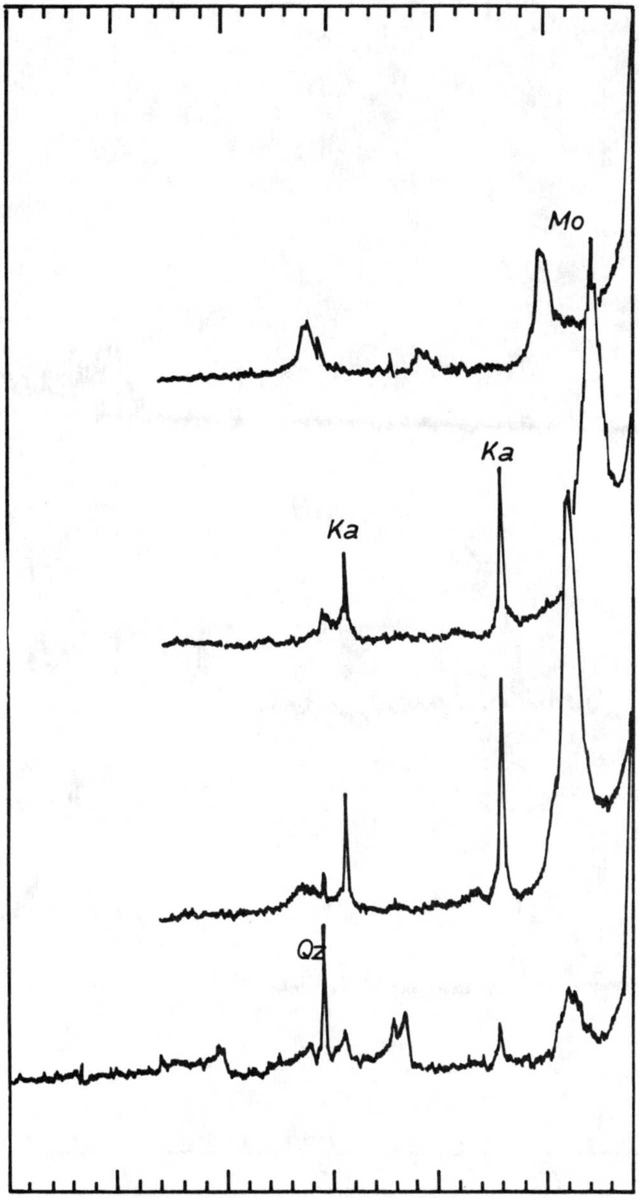

Fig. 9. X-Ray Diffractograms of the Upper Bentonitic Clay Bed, Qasr El-Sagh, Fayum.

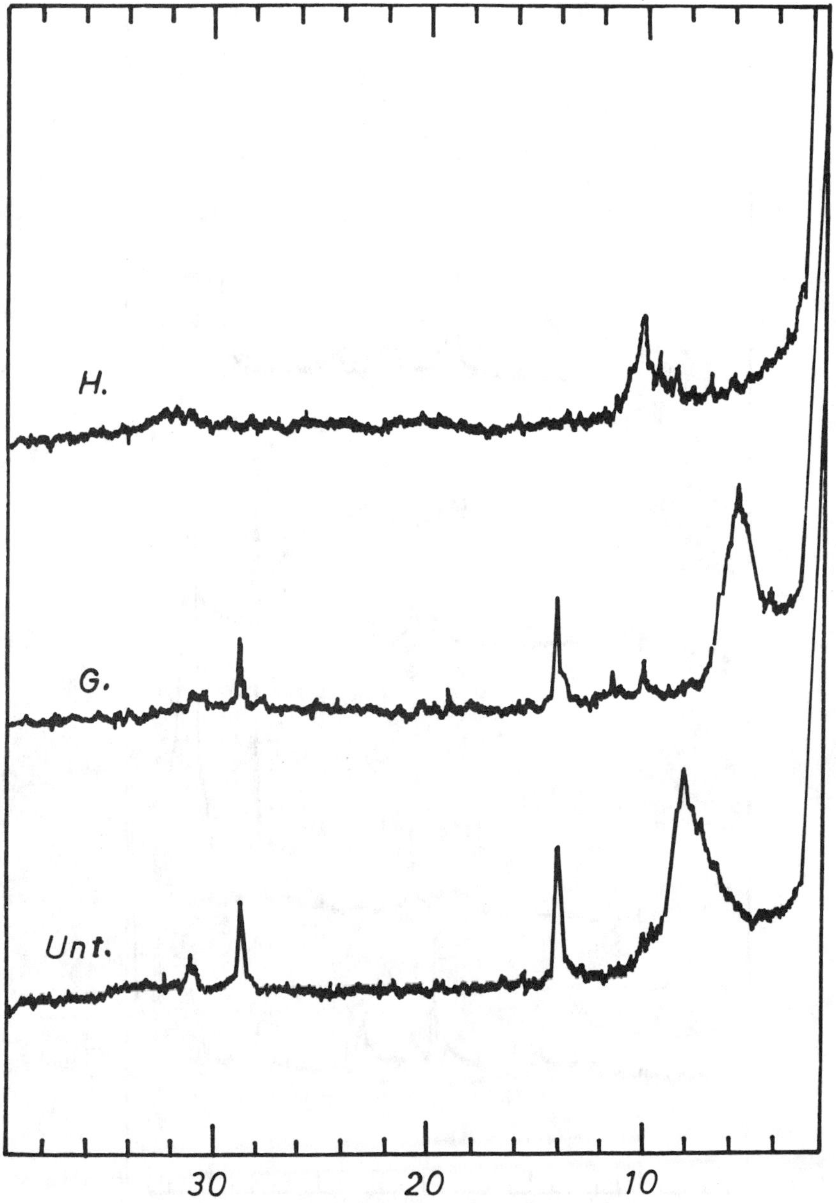

Fig. 10. X-Ray Diffractograms of Bentonitic Clay Sample, Km 38, Cairo-Alexandria Desert Road.

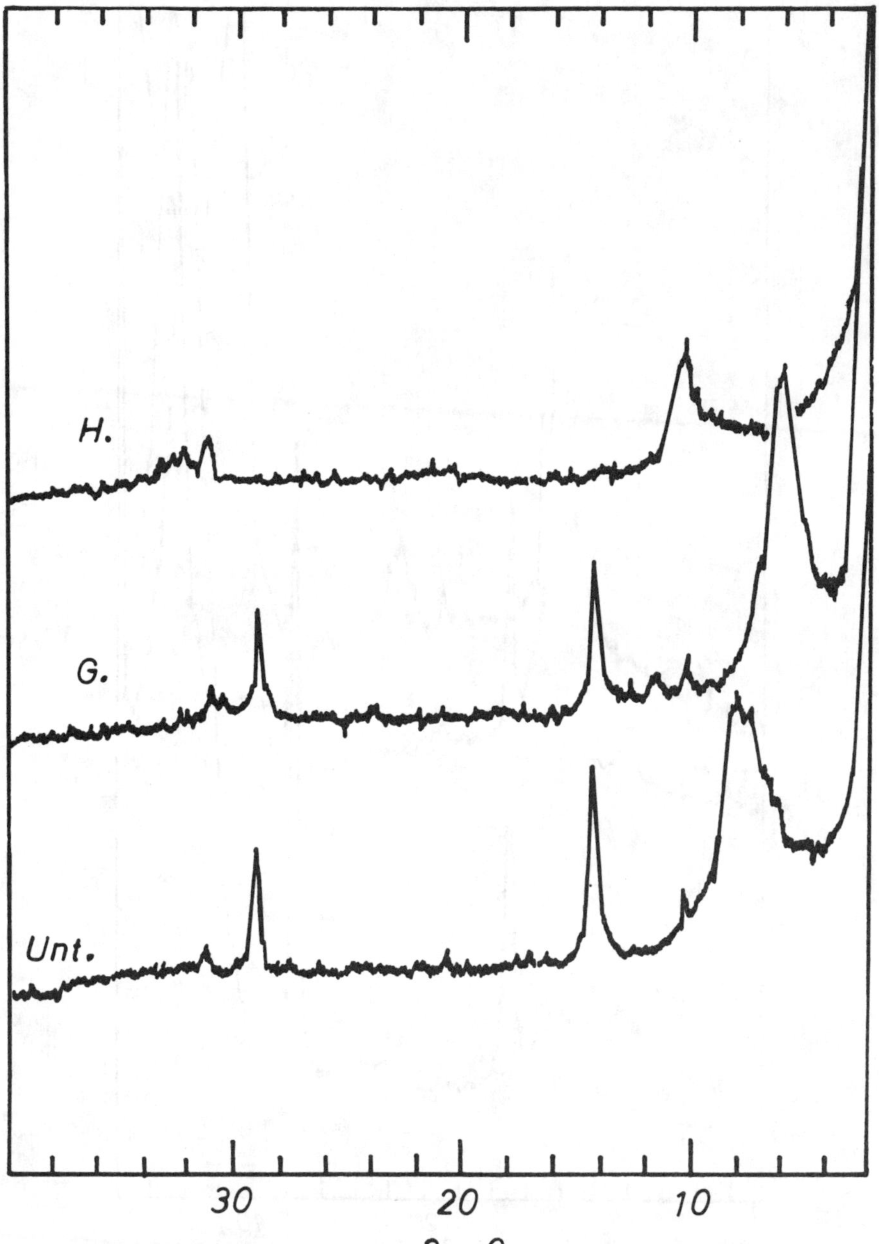

Fig. 11. X-Ray Diffractograms of Bentonitic Clay Sample, Km 60, Cairo-Alexandria Desert Road.

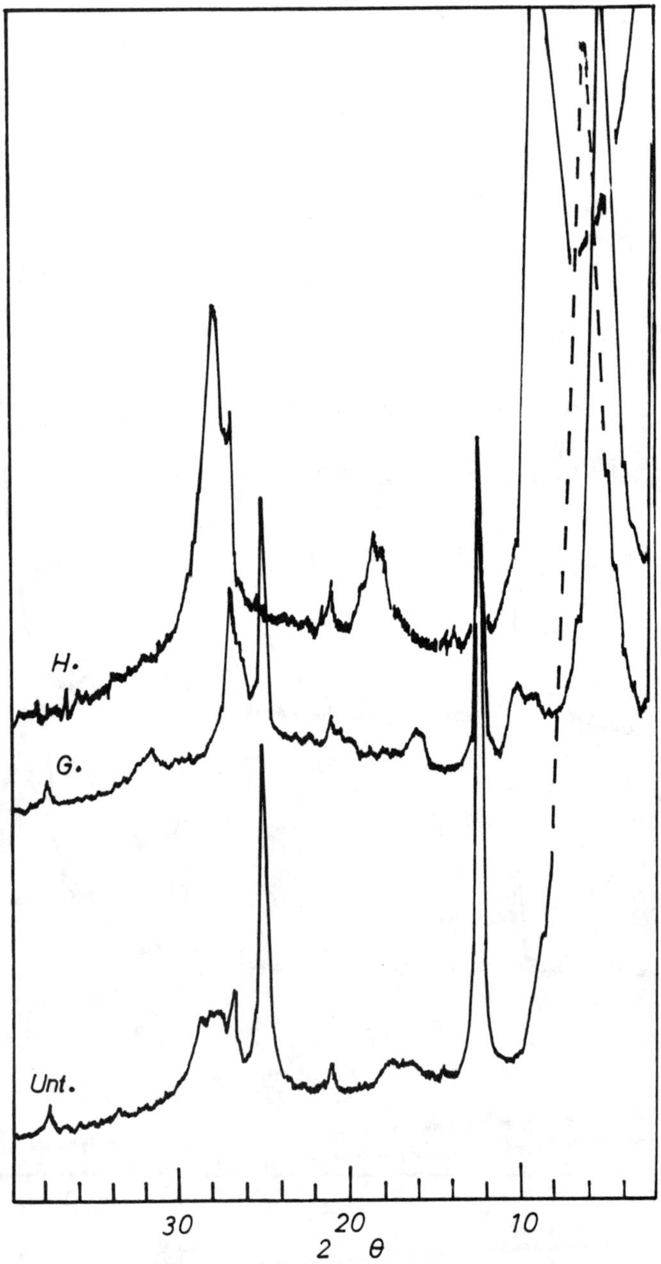

Fig. 12. X-Ray Diffractograms of Bentonitic Clay Sample, Km 25, East-Maadi Road, Qattamia.

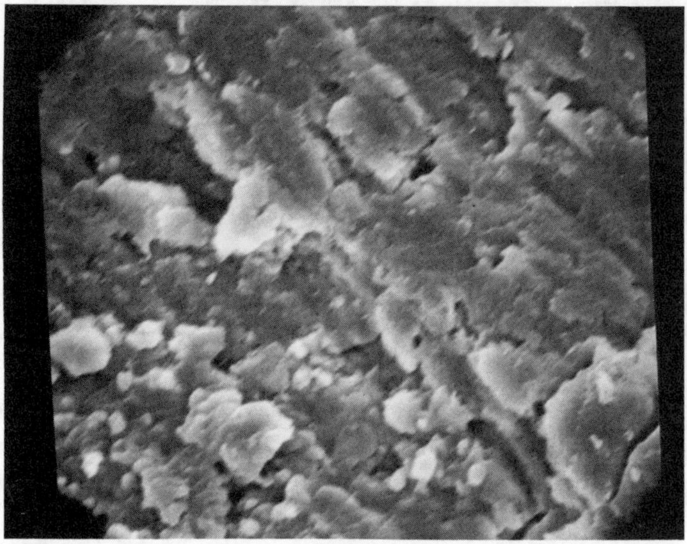

Fig. 13. Electron Photomicrograph Showing Lamellar Montmorillonite Aggregates.

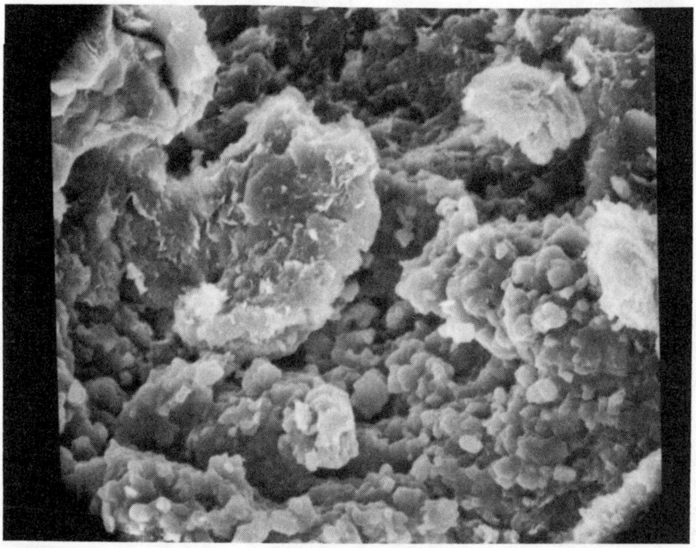

Fig. 14. Electron Photomicrograph Showing Globular Montmorillonite Aggregates.

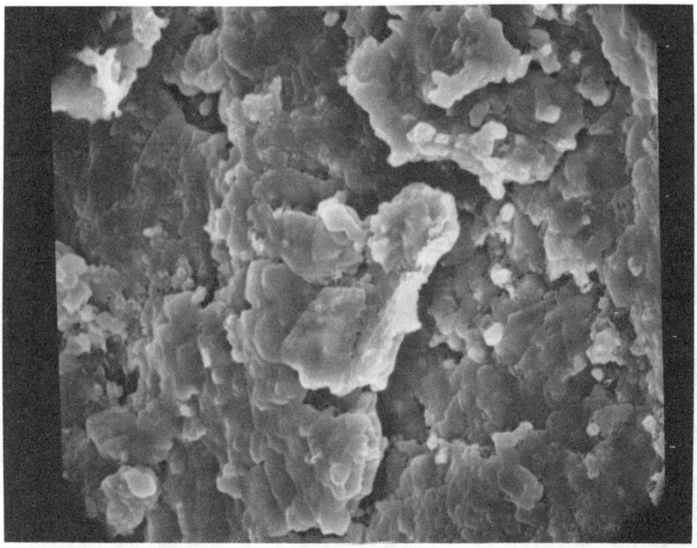

Fig. 15. Electron Photomicrograph Showing Moss-Like Montmorillonite Aggregates.

Relative Proportions of Clay Minerals

The method proposed by Brooks and Ferrel (1970) to quantify clay minerals was adopted, and the peak heights were taken to be equivalent to the relative amounts of the constituent minerals. The intensity measurements of each mineral were expressed as a percentage of the total, and are given in Table VI. With the exception of quartz and feldspars, no minerals other than montmorillonite, kaolinite, montmorillonite-illite mixed layer, and illite were identified in the samples.

From Table VI it can be seen that montmorillonite is the dominant clay mineral in all of the samples. It ranges from about 57-67%, 44-75%, and 45-81% in Qasr El-Sagha, Cairo-Alexandria, and Qattamia deposits, respectively. The kaolinite content ranges from about 14 to 53% in the Qasr El-Sagha deposits, between 14 and 44% in the Cairo-Alexandria Desert road deposits, and from about 18 to 53% in the Qattamia area.

The relative proportion of illite ranges from 0.76 and 12.5% in Cairo-Alexandria and Qattamia clay samples.

Montmorillonite-illite mixed layer clays were identified in the Qattamia deposit. Seven of 10 representative samples contain this type of clay, but the expandable mineral fraction in the mixed layer clay was found to be more than 90%, according to Weaver's curve (Weaver, 1956).

TABLE VI. Relative Proportion of Montmorillonite and Other Clay Minerals in the Samples Examined

Sample	Clay Fraction %	Minerals				Expand. Mineral %	Qtz & Others
		Mo	Ka	Il	Mo-Il		
F1	44.40	57.70	35.00	-	-		7.3
F2	60.70	60.20	36.10	-	-		3.7
F3	60.30	57.80	37.80	-	-		4.4
F4	35.50	67.00	33.00	-	-		-
F5	69.80	60.40	36.00	-	-		3.60
F6	47.80	59.00	34.00	-	-		7.00
C1	41.90	55.90	38.20	-	-		5.90
C2	37.00	51.60	38.70	-	-		9.70
C3	20.40	75.10	13.70	2.5	-		8.70
C4	22.60	57.50	22.50	12.5			7.50
C5	49.40	44.70	39.90	11.0	-		4.4
C6	33.70	50.00	43.50	-	-		6.4
Q1	53.00	-	34.87	0.76	64.37	>90	-
Q2	47.00	62.35	37.65	-	-		-
Q3	58.80	-	39.60	-	60.76	>90	-
Q4	88.10	-	28.80	5.66	65.54	>90	-
Q5	82.90	-	53.48	1.75	44.77	>90	-
Q7	37.40	-	17.94	1.33	80.73	>90	-
Q7a	68.40	-	24.53	-	67.34	>90	8.13
Q8	75.30	-	32.00	3.0	65.00	>90	-
Q9	43.60	72.00	33.00	-	-		-
Q10	67.00	68.55	31.45	-	-		-

Thermal Analyses

Four thermal curves (Fig. 16) were selected to represent the thermal behavior of the three clay deposits. From this figure it can be seen that a large endothermic peak occurs between 100°C and 200°C, accompanied by a loss in weight that ranges between 10% and 14% of the total sample. According to Barshad's tests (1950), this DTA peak is due to the escape of absorbed and interlayer water.

The endothermic "shoulder" at about 280-310°C in the Qasr El-Sagha sample may be due to the loss of water from colloidal constituents such as gels of silica, or to interlayer water loss from clay minerals and/or to dehydroxylation of small amounts of iron hydroxides.

A third endothermic effect in the range 500-700°C is mainly associated with loss of structural hydroxyl ions. Kaolinite dehydroxylates at temperatures ranging from 500 to 700°C depending upon the degree of crystal perfection, while some expanding minerals begin to lose structural OH^- at 500° and 650°C.

The large (more or less symmetric) endothermic peak between 570° and 610°C is caused mainly by the loss of structural OH from montmorillonite and kaolinite (Grim and Bradley, 1940). This reaction is accompanied by a loss in weight ranging from 3 to 7%. In the case of Cairo-Alexandria road samples (km 48) there is a shape change in the endothermic reaction peaks in this temperature range. The endothermic peak at about 600°C may be due to the loss of structural OH from kaolinite, while the "shoulder" at about 670°C may be due to the loss of structural OH from montmorillonite.

The weak endothermic effect that takes place at about 780°C in the case of the Qasr El-Sagha sample may be attributed to the dissociation of small amounts of carbonate minerals.

Most of the clay minerals have an endothermic reaction in the 900-1000°C range. This is often followed by an exotherm that is attributed to recrystallization. The S-shaped endotherm-exotherm peak for montmorillonite appears at 850° to 950°C, and for illite at about 900°C.

The endothermic component of this system is a subject of controversy. It may be attributed either to the destruction of the montmorillonite and illite structures (Grim and Bradley, 1940), or to the loss of hydroxyls present in the silica layer within the tetrahedral configuration of the montmorillonite structure (McConnell, 1950).

The exothermic component, on the other hand, represents the formation of new phases at the expense of both montmorillonite and illite structures. The original composition of such clay minerals determines the type of newly formed phases (Bradley and Grim, 1951).

From Figure 16 it is seen that the endothermic/exothermic "S"-shaped reactions of the bentonitic clay samples have different reaction temperatures. This may be due to the difference in relative proportions of the clay minerals present.

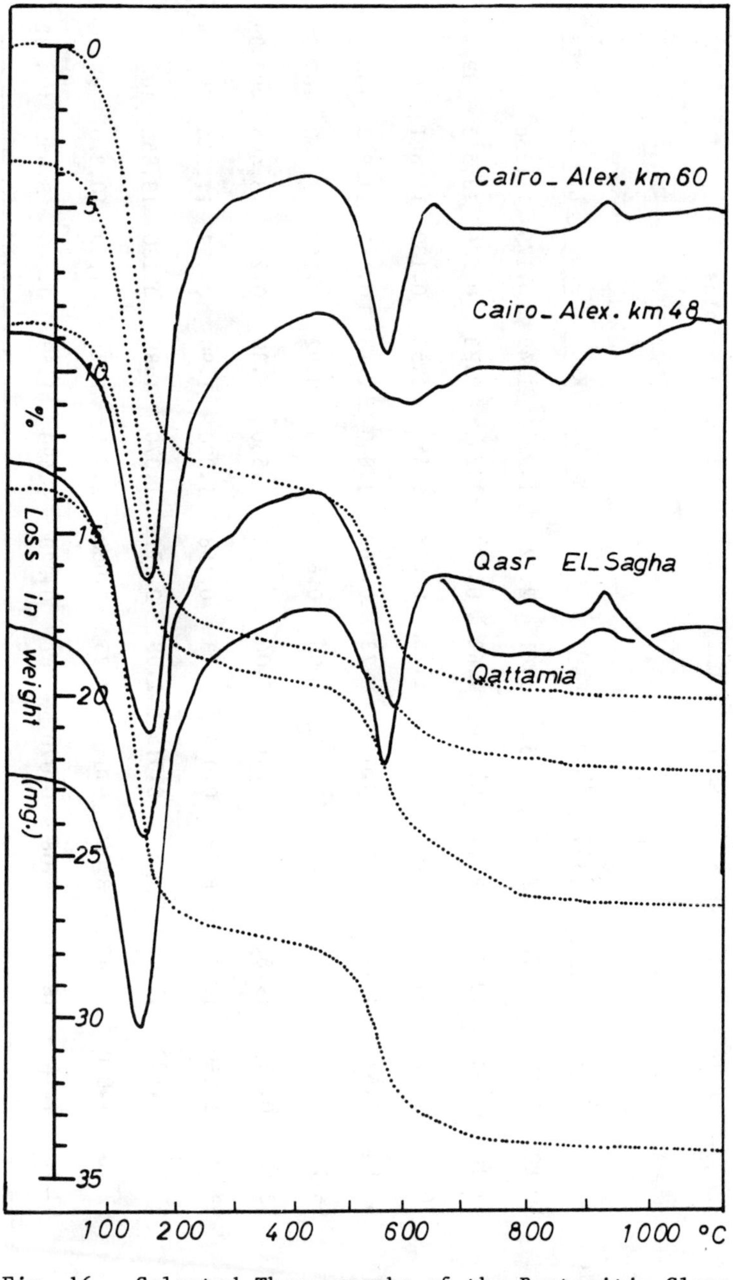

Fig. 16. Selected Thermographs of the Bentonitic Clays.

TABLE VII. Chemical Analysis Data of Bentonitic Clay Samples

Sample	SiO_2	TiO_2	Al_2O_3	Total as Fe_2O_3	MnO	MgO	CaO	Na_2O	K_2O	P_2O_5	L.O.I.	Total
F1	58.80	1.17	14.70	6.47	0.23	1.61	2.72	1.02	1.49	0.175	11,336	99.72
F4	52.10	1.38	19.80	8.44	0.016	1.63	0.327	1.25	0.83	0.206	13,545	99.53
C1	56.00	1.29	21.30	6.06	0.022	1.23	0.058	1.45	1.55	0.130	11,861	100.95
C3	58.80	1.62	15.70	5.78	0.066	2.73	0.760	1.840	1.13	0.152	11,848	100.43
C5	56.74	1.40	22.48	6.78	0.050	0.06	0.650	2.940	0.92	0.140	7,840	100.00
Q4	52.80	0.993	19.80	7.49	0.015	2.06	0.236	1.580	1.30	0.226	13,493	99.99
Q7a	59.00	1.160	17.00	6.48	0.014	1.73	0.158	1.000	1.90	0.152	11,246	99.84
Q7	53.30	1.130	19.00	7.53	0.015	2.08	0.177	1.070	1.88	0.180	13,852	100.21
Q9	52.70	1.010	17.42	8.06	0.040	2.25	3.080	2.710	1.71	0.160	10,370	99.52
Q10	53.16	1.180	18.10	14.08	0.070	1.61	0.220	1.490	1.53	0.240	8.650	100.32

Chemical Analysis

The chemical analysis data for selected samples are listed in Table VII. From the table it is seen that the most common interlayer cations are Na, Mg and K. The variations in basal spacings appear to correlate with the interlayer population.

SUMMARY AND CONCLUSIONS

The clay mineralogy of Qasr El-Sagha and Qattamia bentonitic deposits south of Cairo indicates deposition under shallow marine conditions, whereas those along the Cairo Alexandria Desert road northwest of Cairo are mainly lacustrine in origin.

Sodium montmorillonite is the dominant mineral in the three localities, followed by kaolinite.

Reserve estimates of Qasr El-Sagha deposits are about eight million tons. According to preliminary calculations, reserves of bentonitic clays along the Cairo-Alexandria Desert road are about two million tons. An extensive drilling program is being conducted to estimate the reserves along the entire road where the reserves are expected to surpass this figure by a factor of three or more.

Preliminary tests show that the Qasr El-Sagha deposits may be suitable for foundry use after mechanically removing the gypsum. Test work also indicates that the deposits along the Cairo-Alexandria Desert road can be used as drilling mud. Standard dry processing of this deposit is all that is necessary to produce a good quality bentonite.

The Qattamia bentonitic clay deposits are now being used for agricultural purposes. It is being mixed with desert sand to make productive soils that will retain water and plant food.

REFERENCES

Barshad, I., 1950, The Effect of the Interlayer Cations of the Expansion of Mica-Type of the Crystal Lattice, Am. Min. V, 35, pp. 225-238.

Basta, E. Z., El-Kadi, M. M., and Madsoud, M. A., 1971, Mineralogy of Some Bentonitic Clays from Fayum, Cairo University Faculty of Science Bulletin, V. 43, pp. 271-284.

Bradley, W. F. and Grim, R. E., 1951, High Temperature Thermal Effects of Clay and Related Materials, Am. Min., V. 36, pp. 182-201.

Brooks, R. A. and Ferrel, R. E., 1970, The Laterial Distribution of Clay Minerals in Lakes Pontchartrain and Marepas, Louisiana, Journal Sed. Petrology, V. 40, pp. 855-863.

Grim, R. E. and Bradley, W. F., 1940, Investigation of Effect of Heat on Clay Minerals, Illite and Montmorillonite, Journal Am. Ceram. Soc., V. 23, pp. 242-248.

Hegab, O. A., 1974, Clay Mineralogy and Petrography of Some Clay Deposits in El-Fayum Area, Egypt. Ph.D. Thesis, Submitted to Ain Shams University, Cairo, Egypt.

Hendricks, S. G., Nelson, R. A., and Alexander, L. T., 1940, Hydration Mechanism of the Clay Mineral Montmorillonite Saturated with Various Cations, Journal Am. Chem. Soc., V. 62, pp. 1457-1464.

Kilmer, V. J. and Alexander, L. T., 1949, Methods of Making Mechanical Analysis of Soils, Soil Science, R. 63, pp. 15-24.

McConnell, D., 1950, The Crystal Chemistry of Montmorillonite, Am. Min., V. 35, pp. 166-172.

Weaver, C. E., 1956, The Distribution and Identification of Mixed-Layer Clays in Sedimentary Rocks, Am. Min., V. 41, pp. 202-221.

Chapter 11

KERR ON SENSITIVE SOILS AND QUICKCLAYS

by Ian J. Smalley

Earth Sciences Department
University of Waterloo
Waterloo, Ontario, Canada

ABSTRACT

"Quickclay is an extreme case; it is by far the most mobile of all the common solid materials on the earth's surface. It has both a high water content and a mineral texture that allows it to flow with the utmost ease. A mass of quickclay that has lain undisturbed for thousands of years can be jarred into motion by any sudden shock." The quotation is from the paper by Kerr called, "Quickclay," which was published in Scientific American in 1963. This was probably the most widely read paper on quickclay ever published. It was reprinted by Fred Betz in his Benchmark collection on Environmental Geology (BM 25).

A quickclay is an extremely sensitive soil. The sensitivity parameter is produced by dividing the undisturbed shear strength by the remoulded shear strength. If a value greater than 50 is obtained, then the soil can be classified as a quickclay. True quickclays are characterized by a very low remoulded strength of the order of 0.5 kPa. The classic quickclays are found in Scandinavia and Eastern Canada, but the 1964 Alaska earthquake revealed a considerable quickclay deposit underlying the town of Anchorage. This was discussed in a classic paper by Kerr and Drew in Engineering Geology in 1967.

The Kerr views on quickclays were perhaps most clearly expressed in the paper with Liebling in Geol. Soc. Amer. Bull. in 1965. They suggested that the major critical conditions which contribute to quickclay movement are: substantial quantities of layer lattice silicates of colloidal size (about 40% or more by dry weight), substantial pore water (44%), a reduction in electrolyte concentration below 5 g/l of salt, and addition of a dispersant. There is still discussion and disagreement about the nature of quickclays and the mechanisms of quickclay slides; the "clay mineral"-based views favored by Kerr still find support, although an alternative "primary mineral" view suggests that it is the finely ground primary minerals (quartz, feldspars, etc.) which determine the quickclay properties and that a very small amount of clay mineral material is required.

REFERENCES

Drew, I. M. and Kerr, P. F., "Quickclay Slides in the U.S.A.," 1967, Engineering Geology, Recent Quickclay Studies, 5, Elsevier Publishing Company, Amsterdam.

Kerr, P. F., 1963, "Quickclay," Scientific American, 209, pp. 132-138.

Liebling, R. S. and Kerr, P. F., 1965, "Observations on Quickclays," Geol. Soc. Am. Bull., 76, pp. 853-878.

GEOCHEMISTRY, ALTERATION AND ORE MICROSCOPY

Chapter 12

COMPARISON OF THE GEOCHEMISTRY OF LIMESTONES AND DOLOSTONES ADJACENT TO MINERALIZED BRECCIAS (COY MINE, MASCOT-JEFFERSON CITY DISTRICT) AND NON-MINERALIZED BRECCIAS (THORN HILL SECTION) IN EAST TENNESSEE

by Otto C. Kopp, Thomas B. Crattie and Dennis R. Owen

Department of Geological Sciences, University of Tennessee, Knoxville, Tennessee 37916

ABSTRACT

Whole-rock analyses of limestones, fine-grained dolostones and coarser-grained dolostones adjacent to mineralized breccias (Coy mine, Mascot-Jefferson City district) were compared with similar lithologies adjacent to apparently non-mineralized breccias at the new Thorn Hill section located approximately 30 km north of the Coy mine. Specimens were analyzed for Ca, Mg, Si, Al, K, Fe, Mn, Sr, Br, Cl, S, Ti and P using X-ray fluorescence analysis. The concentrations of the major and most minor elements for Thorn Hill specimens fall within the range of concentrations for Coy mine specimens; however, significant differences (probability < 0.05) were noted for Fe, Cl and Br for fine-grained dolostones and for MgO, Fe and Mn for the coarser-grained dolostones. No significant differences were detected for the limestones.

INTRODUCTION

Previous Studies

The literature is replete with studies of Mississippi Valley-type mineralization and much attention has been devoted to the nature and origin of these deposits in Tennessee. In East Tennessee, significant mineralization is found mainly in the upper part of the Kingsport and lower part of the Mascot Formations, both of Ordovician age. The mineralogy of these deposits is rather simple (dolomite, calcite, various forms of silica, clays, sphalerite, minor pyrite, galena, barite, fluorite, etc.) and they typically occur in association with breccia bodies and in bedded structures in certain limestone beds in the Kingsport. In breccia bodies, the ore minerals and carbonate gangue fill the interfragmental space. Fluid inclusion evidence (Roedder, 1971; Caless, 1983; etc.) suggests that the ore fluid(s) were derived from basinal brines and had homogenization temperatures ranging from 70-200° C.

It is not the intention of this paper to review the many hypotheses which have been proposed to explain the origin of these deposits. The interested reader is directed to review articles by Hoagland et al. (1965), Crawford and Hoagland (1968), McCormick et al., (1971), Hoagland (1976), Churnet and Misra (1983), McCormick et al., (1985), and others.

In a recent study in Central Tennessee, Ferguson (1981) noted that some elements in the gangue carbonates and adjacent host breccia fragments appear to have high correlation coefficients. A question arises as to just how much influence the country rock and ore fluids bear upon each other. One possibility is that the fluids which brought in the ore minerals were also responsible for altering the minor and trace element compositions of adjacent wall rocks. Alternatively, the host breccias through which ore fluids passed may have modified those fluids through processes such as partial solution, diffusion, etc. It is also possible that the ore fluids and host breccias modified each other to some degree.

Relationships between chemical composition and lithologic character at the Coy mine and other locations in East Tennessee have been discussed by West (1974), Snyder (1975), Churnet and Misra (1981), Moss (1982), Churnet et al., (1983), Crattie (manuscript in preparation), and others. In an attempt to determine the extent to which rocks adjacent to mineralized breccias may have been influenced by these fluids, Crattie (manuscript in preparation) determined the lithologic and chemical character of samples from a wide vertical range of lithologies within mineralized sections of the Kingsport Formation in the Coy Mine. Samples were taken from several stopes (Table 1) at distances from less than one meter to as much as twenty meters away from breccia bodies to determine the nature and extent of compositional changes at specific stratigraphic levels.

In the Coy mine, Crattie (manuscript in preparation) recognized three dominant lithologic types: limestones (Ls) (micrites, biomicrites, intrapelsparites, etc.) and their partially dolomitized equivalents; xenotopic,

TABLE 1. Sample Locations

Lithology	Sample Numbers
Coy mine	
Limestones:	Stope 9-50: C1A, C2B, C3B Stope 25-56: C10A, C13B, C18, C19
XFd Dolostones:	Stope 9-50: C3A, C3C, C4B, C5B Stope 23-49: C20A, C23B Stope 25-56: C11B, C12A, C16B
XHMCd Dolostones:	Stope 9-50: C4A Stope 23-49: C20B Stope 25-41: C22A2, C23A Stope 25-56: C11A, C12B, C13A
Thorn Hill	
Limestones:	K07 (9.5)[a], K010 (15.0), K011 (17.0), K012 (20.0), K013 (23.0), K014 (26.0)
XFd Dolostones:	K01 (0.5), K03 (4.0), K05 (6.0), K06 (8.5), K08 (12.0), K016 (31.5)
XHMCd Dolostones:	K021 (+0.5), K02 (1.5), K04 (5.0), K022 (5.0), K015 (28.0), K017 (35.0), K018 (40.5), K019 (44.5), K020 (49.0)

[a] Distances in meters below Kingsport-Mascot contact, except sample K021 which was taken 0.5 m above the contact.

fine-grained dolostones (XFd); and xenotopic to hypidiotopic, medium- to coarse-grained dolostones (XHMCd). High correlation coefficients for limestones were determined for elemental pairs such as Fe:Mg, Br:Fe, Fe:Mn, etc. An unusual relationship between Br and Cl and lithologic type was discovered. Relatively high Cl levels were anticipated for the limestones as were correspondingly low Br levels for those rocks; however, although higher Br concentrations were anticipated for dolostones (especially the XFd which are thought by most workers to be penecontemporaneous dolostones formed in intertidal to supratidal environments), the low Cl levels found in the dolostones were not expected.

Purpose of the Present Study

The purpose of this study is not simply to compare the relationships (or lack thereof) between composition and lithologic character, such as the higher contents of Fe and Mn in dolostones and dolomitized limestones, but rather to compare the compositions of similar lithologies (limestones and dolostones of the Kingsport Formation) adjacent to mineralized breccias and non-mineralized breccias to see whether the same geochemical relationships exist. If so, this might suggest that the ore fluids and host breccias had little influence on each other. On the other hand, if there were some chemical differences, then perhaps this information might help to explain the origin of these deposits, or be useful as an exploration tool.

EXPERIMENTAL

Samples

The samples used in the present study came from the Coy mine, located in the Mascot-Jefferson City district (approximately 45 km northeast of Knoxville) and from the new Thorn Hill section (located approximately 60 km northeast of Knoxville) in the Copper Ridge district (Figure 1). The Coy mine and Thorn Hill section are separated by several northeast-trending thrust faults including the Saltville thrust. Although no sphalerite mineralization was observed at the Thorn Hill roadcut, sphalerite-bearing breccias do occur in the Idol mine (approximately 2 km to the northeast) and the Flat Gap mine (approximately 20 km to the northeast).

Ideally, a non-mineralized breccia in the Mascot-Jefferson City district might have been more suitable; however, the Copper Ridge and Mascot-Jefferson

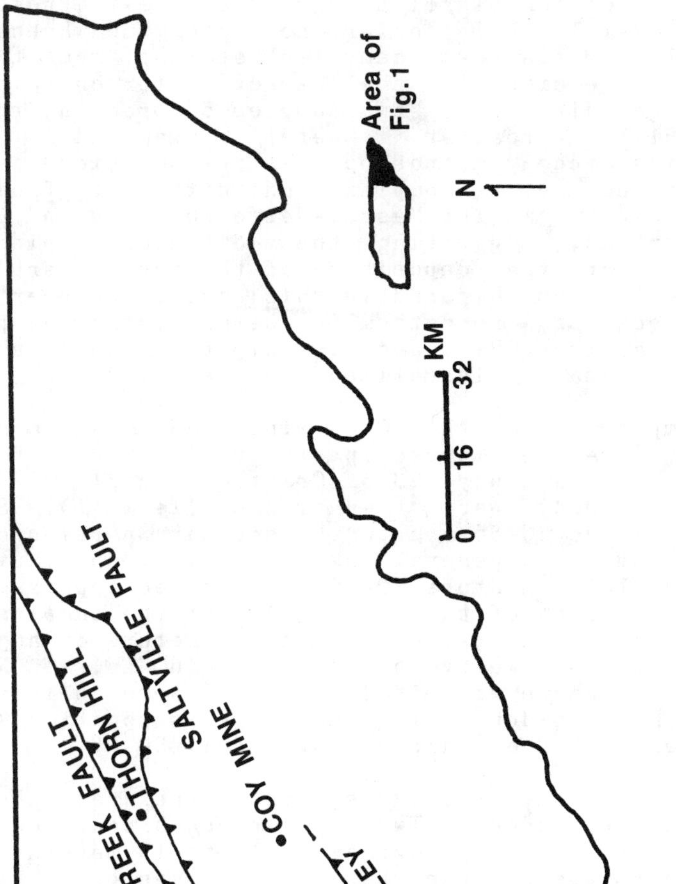

Fig. 1. Location of Coy mine and Thorn Hill section in Tennessee.

City districts appear to have been mineralized at the same time and by the same ore fluid(s) based on a study of the microstratigraphic zonation pattern in the gangue carbonates (Ebers, 1976; Ebers and Kopp, 1979) and the Thorn Hill section is relatively fresh, well exposed and easily accessible. In order to determine whether the Thorn Hill breccia was cemented at the same time as mineralized breccias in the Mascot-Jefferson City and Copper Ridge districts, two samples of breccia from the Thorn Hill roadcut were examined using cathodoluminescence microscopy. The zonation pattern appears to be quite complex and different from the pattern observed in the Mascot-Jefferson City and Copper Ridge districts, suggesting that different fluids were responsible for the deposition of the gangue carbonates in the two sites compared in this study. Further study will be necessary to determine the microstratigraphy of the gangue at Thorn Hill and its temporal relationship to the gangue of nearby mineralized breccias.

The samples from the Coy mine had been analyzed previously (Crattie, manuscript in preparation). For the purposes of this study, 23 of Crattie's samples (7 Ls, 9 XFd and 7 XHMCd) were re-analyzed (Table 1). Stopes 23-49, 25-41 and 25-56 are in the southern portion of the mine and lie on a general east-west line, approximately 550 meters long. Stope 9-50 is located approximately 1000 meters north of this line. All of the samples were collected from within less than 20 meters of adjacent breccia bodies. Twenty-one samples (6 Ls, 6 XFd, 8 XHMCd, and 1 gangue dolomite) (Table 1) were collected at the Thorn Hill roadcut. The beds at this location strike approximately N65^{o}E and dip approximately 30°SE.

Numerous joints, fractures, and brecciated zones are exposed in the roadcut. Two major breccia zones occur in the intervals from approximately 10 to 15 meters and 25 to 35 meters below the top of the Kingsport. All the samples were collected within less than 20 meters of a breccia zone. The samples were collected near road level and all distances were measured normal to the bedding.

Sample Preparation

Selected fragments of each whole-rock sample were crushed to pea-size fragments and examined visually. Fragments which contained fracture fillings or showed any signs of weathering were discarded (except for one sample of gangue dolomite in a breccia which was analysed). No attempt was made to avoid or remove chert, clays, etc. The retained material was washed with deionized water and oven dried. Subsequently, all material was handled with clean gloves to avoid contamination with chlorides. Several grams (minimum 6 g) were ground in a Shatterbox for six minutes using a tungsten carbide mill. The resulting powders were pressed into boric acid-backed briquettes using 103 MPa (15,000 psi) and stored in a dessicator prior to analysis.

Analytical Conditions

The samples from both localities were analyzed in random order using tube-excited fluorescence analysis (EG&G ORTEC TEFA III). The analytical conditions are given in Table 2. No attempt was made to analyze for zinc or other several other trace elements at this time. The limestone and dolostone standards were provided by EG&G ORTEC. The chlorine standards were prepared by spiking one of the dolomite standards (NBS No. 88a) using reagent grade KCl. The bromine standards were prepared by analyzing four of the Coy mine samples by neutron activation analysis and then re-analyzing these samples using X-ray fluorescence. Replicate analyses were performed on several samples.

RESULTS AND DISCUSSION

Analytical Results

The analytical results are given in Table 3 (Ls), Table 4 (XFd) and Table 5 (XHMCd). The results for Ca, Mg and Si are given in weight percent oxides, while the other elements are given in ppm. The analytical results for Al, P, K and Ti were not used in the present study. The results of t-tests of the data for those elements for which significant (probability < 0.05) differences were determined are given in Table 6.

Limestones

The chemical data for limestones from the Coy mine and Thorn Hill are presented in Table 3. Limestones from both locations have remarkably similar trace element compositions, in spite of varying degrees of dolomitization and silicification. Although the data are somewhat limited in this study (7 samples from Coy mine and 6 from Thorn Hill), the two data sets are statistically indistinguishable on the basis of their CaO, MgO, SiO_2, Fe, Mn, Sr, and Cl concentrations.

The mean Br and S contents of the Coy mine samples

TABLE 2. Analytical Conditions for TEFA Analyses

No.	KV	UA	Energy Range	Filter	Anode	Elements Analyzed
1	10	100	0-10 Kev	Open	Mo	Mg, Al, Si, P, Ca, Ti, Mn, Fe
2	10	80	0-10 Kev	Open	W	S, Cl
3	45	222	0-20 Kev	Sn	W	Br, Sr

Notes: All analyses were performed under vacuum and with a 200 second live time.
Both Beer's and stripping corrections were applied, as necessary.

TABLE 3. Analytical Results for Limestones

Sample	CaO	MgO	SiO_2	Fe	Mn	Sr	Cl	Br	S
Coy Mine									
C1A	49.32	0.57	10.35	1353	50	223	3186	-	600
C2B	45.66	2.44	11.03	580	121	303	2944	2	139
	(46.88	2.52	11.24	852	166	286	2909	2	73)
C3B	31.39	9.28	24.97	1526	335	164	632	8	3869
C10A	44.49	1.46	12.41	674	95	191	2617	-	1235
C13B	39.65	10.03	4.12	3228	516	168	2100	7	83
C18	54.86	0.43	1.88	637	1	240	4564	-	137
C19	52.41	0.11	2.14	288	-	242	3823	-	145
Mean:	45.40	3.47	9.56	1184	160	219	2838	2	887
S.D.:	8.00	4.30	8.08	1004	194	49	1261	4	1379
Thorn Hill									
K07	51.65	0.64	6.67	582	47	225	3553	-	175
K010	43.95	2.05	19.21	588	172	174	2033	-	149
K011	40.15	9.63	3.95	3489	533	195	2432	1	37
K012	54.00	0.53	2.71	741	17	309	3903	-	168
K013	50.10	0.43	6.96	1136	9	246	3251	-	183
	(49.86	0.51	7.10	1112	-	248	3265	-	135)
K014	45.90	3.54	7.20	1480	185	278	2560	-	114
Mean:	47.63	2.80	7.78	1336	161	238	2955	-	138
S.D.:	5.20	3.56	5.89	1112	198	51	724	-	55

Note: Replicate analyses enclosed in ( ). These values were not included in the calculations of mean and standard deviation.
Concentrations in weight % for CaO, MgO and SiO_2; all others in ppm.

appear to be slightly elevated relative to those from Thorn Hill, but they cannot be distinguished by the statistical treatment used. The higher Br contents are associated with higher MgO concentrations (eg. sample nos. C3B and C13B) and most of the samples from the Coy mine have relatively low, <200 ppm, S contents.

Since the Coy mine samples came from within the mine, it is possible that traces of sulfides were included with some of the limestone samples. That the elevated Br contents in some Coy mine samples are related to dolomitization is also suggested by the presence of detectable Br in sample K011 (the most dolomitic of the Thorn Hill samples). The presence of higher concentrations of Br in dolostones is discussed below.

Xenotopic, Fine-grained Dolostones

Compositional data for XFd samples from the Coy mine and Thorn Hill are given in Table 4. The weight percentages of CaO and MgO are essentially identical for samples from both areas. Note that silica contents in these rocks can be quite variable. Most workers consider these rocks to be of "primary" origin or nearly so. The concentrations of Sr and S are comparable. However, the concentrations of Fe, Cl and Br appear to be greater in samples from the Coy mine. (The concentration of Mn may also be slightly elevated in Coy mine samples.) The results of the t-tests are presented in Table 6.

One sample (K09) of gangue was taken from cm-sized fractures within sample K08. It was essentially free of silica and S, but had higher contents of Fe and Mn and a relatively high content of Cl compared with the enclosing fine-grained dolostone.

Xenotopic-Hypidiotopic, Medium- to Coarse-Grained Dolostones

The compositions of XHMCd rocks (Table 5) from both areas are similar with respect to their CaO and MgO contents. The range of concentration of silica and its mean value appear greater for the samples from the Coy mine, but this may be the result of wide ranges in the amounts of silica present in these rocks. Silica may be present in several forms, including quartz sand and silt, clay minerals, primary and secondary cherts, etc. The samples from the Coy mine have significantly greater concentrations of MgO, Fe, and Mn (Table 6).

Considering dolostones from both the mineralized and

TABLE 4. Analytical Results for XFd Dolostones

Sample	CaO	MgO	SiO_2	Fe	Mn	Sr	Cl	Br	S
<u>Coy Mine</u>									
C3A	30.36	22.15	1.43	5323	287	139	752	9	93
C3C	30.24	21.98	1.46	4799	251	129	941	12	752
C4B	29.32	21.66	4.75	4640	370	136	745	11	284
C5B	29.53	22.46	2.55	4703	366	110	853	7	223
C11B	30.67	19.19	5.48	3440	223	222	1105	5	932
C12A	30.17	20.77	3.53	4293	524	177	837	13	499
	(30.68	20.25	3.53	4664	573	175	821	11	423)
C16B	29.28	22.08	1.74	3661	239	130	123	3	574
C20A	29.54	20.13	4.16	3314	252	208	412	8	1201
C23B	29.32	23.04	1.89	2040	193	132	104	6	765
Mean:	29.83	21.50	3.00	4024	301	154	652	8	591
S.D.:	0.53	1.23	1.53	1007	103	39	357	3	359
<u>Thorn Hill</u>									
K01	30.16	20.75	3.42	1227	209	192	526	6	258
K03	30.17	21.79	1.82	1076	273	178	440	2	-
	(30.03	22.30	1.79	1220	239	181	304	5	-)
K05	27.93	21.55	8.75	2806	242	147	-	4	780

TABLE 4. Analytical Results for XFd Dolostones, Continued

Sample	CaO	MgO	SiO_2	Fe	Mn	Sr	Cl	Br	S
Thorn Hill, continued									
K06	29.65	21.82	4.03	1476	234	182	86	5	310
K08	28.26	22.38	8.62	2820	216	171	-	5	1128
K016	29.16	22.14	4.06	883	168	177	60	2	52
Mean:	29.22	21.74	5.12	1715	224	175	185	4	421
S.D.:	0.96	0.57	2.88	873	35	15	235	2	442
Gangue in K08									
K09	30.62	21.89	-	7174	420	131	422	4	-

Note: Replicate analyses enclosed in ( ). These values were not included in the calculations of mean and standard deviation.
Concentrations in weight % for CaO, MgO and SiO_2; all others in ppm.

TABLE 5. Analytical Results for XHMCd Dolostones

Sample	CaO	MgO	SiO_2	Fe	Mn	Sr	Cl	Br	S
Coy Mine									
C4A	23.30	19.11	41.35	4150	665	119	-	8	-
C11A	27.60	18.27	17.27	4725	660	148	129	7	668
	(28.15	18.63	17.94	5447	740	146	374	7	736)
C12B	28.60	19.78	12.84	7238	1026	140	478	13	338
C13A	32.33	18.82	3.07	6265	985	154	1166	15	50
C20B	27.62	19.24	18.83	5194	869	134	67	8	32
C22A2	28.68	21.80	8.91	5111	713	134	191	9	-
C23A	28.01	21.55	15.38	5908	817	122	-	7	-
Mean:	28.02	19.80	16.81	5513	819	136	290	10	155
S.D.:	3.65	1.36	12.01	1036	149	13	420	3	257
Thorn Hill									
K02	29.57	21.60	5.10	2301	351	146	658	9	157
K04	29.43	21.69	3.94	5599	688	123	591	9	31
K015	29.34	22.16	4.36	1827	258	155	-	-	-
K017	29.23	22.73	7.90	2558	340	120	146	9	-
K018	28.59	21.04	12.84	2598	289	204	421	11	477

TABLE 5. Analytical Results for XHMCd Dolostones, Continued

Sample	CaO	MgO	SiO_2	Fe	Mn	Sr	Cl	Br	S
Thorn Hill, continued									
K019	29.97	21.20	5.27	6344	789	128	393	8	-
K021	27.33	19.42	19.85	3177	338	172	35	8	199
	(26.96	19.64	19.42	3048	294	171	-	6	112)
K022	27.57	21.54	9.13	2774	218	143	357	13	381
	(27.78	21.64	9.35	2958	250	141	327	9	377)
Mean:	28.88	21.42	8.55	3397	409	149	325	8	156
S.D.:	0.96	0.97	5.46	1647	210	28	244	4	187

Note: Replicate analyses enclosed in ( ). These values were not included in the calculations of mean and standard deviation.
Concentrations in weight % for CaO, MgO and SiO_2; all others in ppm.

TABLE 6. Results of Statistical Analyses

Lithology	Location	Element	Mean	S.D.	t-value and Probability
XFd	Coy mine	Fe	4024	1007	
	Thorn Hill	Fe	1715	873	t(13) = 4.575, P < .001
	Coy mine	Cl	652	357	
	Thorn Hill	Cl	185	235	t(13) = 2.808, P = .0143
	Coy mine	Br	8	3	
	Thorn Hill	Br	4	2	t(13) = 2.838, P = .0135
XHMCd	Coy mine	MgO	19.80	1.36	
	Thorn Hill	MgO	21.42	0.97	t(13) = -2.692, P = .0177
	Coy mine	Fe	5513	1036	
	Thorn Hill	Fe	3397	1647	t(13) = 2.924, P = .0115
	Coy mine	Mn	819	149	
	Thorn Hill	Mn	409	210	t(13) = 4.299, P = .001

non-mineralized areas, the concentrations of Fe and Mn appear to be lowest in the XFd samples from Thorn Hill and greatest in the XHMCd samples from the Coy mine.

DISCUSSION AND CONCLUSIONS

It is important to keep in mind that the comparisons made above are based on relatively limited sampling (23 samples from the Coy mine and 20 samples from the Thorn Hill exposure). Clearly, more work is needed in these and other areas to better define the similarities and differences that may exist for carbonate rocks in mineralized and non-mineralized areas. A study of the geochemical character of limestones and dolostones of the Kingsport Formation in areas which lack brecciation may provide useful background information concerning the levels of minor and trace elements that occur in these lithologies as a result of depositional and early diagenetic processes.

Several tentative conclusions can be drawn from the present study:

1. The limestones from both areas range widely in their major element compositions and they could not be distinguished on the basis of their Fe, Mn or Sr concentrations. The greatest changes in these trace element concentrations appear to be associated with dolomitization. Although the Br and S contents of the limestones at the Coy mine appear to be greater than those at Thorn Hill, they were not distinguishable at the 95% confidence level.

2. Fe and Mn (and to a lesser degree MgO, Cl and Br) concentrations differ in XFd and XHMCd in the mineralized and non-mineralized areas. The mean concentration of Fe in Coy mine XFd is >2X the Fe concentration in Thorn Hill XFd. The mean concentrations of Fe and Mn in Coy mine XHMCd are nearly 2X the concentrations in corresponding samples from Thorn Hill. The mean concentrations of Cl and Br in Coy mine XFd are respectively 3X and 2X the mean concentrations at Thorn Hill.

3. As observed by Crattie for the Coy mine samples (manuscript in preparation), the highest Cl concentrations and lowest Br concentrations occur in the limestones from both areas. The Cl concentrations are lower and Br concentrations higher in the dolostones in both areas. An increase in Br concentration might be anticipated to result from any process that increases the salinity of a brine (Rittenhouse, 1967), whether

concentrated by evaporation or by some subsurface mechanism (Graf, 1982); however, it is difficult to explain the concomitant, very large decrease in Cl in the dolostones (note that in some samples Cl was below the detection limit, approximately 35 ppm). Nor does a fresh water-saline water mixing model seem adequate to explain the inverse relationship between Cl and Br in the limestones and dolostones in these two areas. At present, we do not mean to imply any genetic relationship between dolomitization and the levels of Cl and Br detected; however, if further study confirms the concentration levels determined for Cl and Br (especially for the low levels of Cl in some dolostones) in this study and their inverse relationship it may be necessary to take a new look at the process(es) responsible for dolomitization and mineralization.

4. Based on the samples analyzed, the most distinctive chemical characteristics of the rocks in the mineralized area are their higher concentrations of Fe and Mn. Note that Viets et al (1983) observed decreasing Fe and Mn contents in the carbonate phase of the Bonneterre away from the Lamotte-Bonneterre contact. The Fe and Mn contents were said to be anomalous in the Virburnum Trend.

ACKNOWLEDGMENTS

The authors wish to thank the staff of EG&G ORTEC, especially Nancy Jacobus, for their assistance in analyzing the samples used in this study. We appreciate the constructive comments offered by Stuart Maher (Kenwill, Inc.) and Del Harper (ASARCO, Inc.). We also wish to acknowledge the Professors' Honors Fund, Department of Geological Sciences, University of Tennessee, for partial support of this research.

REFERENCES

Caless, J. R., 1983, "Geology, Paragenesis and Geochemistry of Sphalerite Mineralization at the Young Mine, Mascot-Jefferson City District, East Tennessee," (Unpublished Master's Thesis, Virginia Polytechnic Institute and State University).

Churnet, H. G. and Misra, K. C., 1981, "Genetic Implication of the Trace Element Distribution Pattern in Upper Knox Carbonate Rocks, Copper Ridge District, East Tennessee," Sedimentary Geology, Vol. 38, pp. 279-300.

______________________________, 1983, "Sphalerite Mineralization and Its Relationship to Carbonate Facies Boundaries, Copper Ridge District, East Tennessee," Proceedings, International Conference on Mississippi Valley-Type Lead-Zinc Deposits, University of Missouri, Rolla, pp. 360-372.

Crattie, T. B., (Manuscript in Preparation), "A Petrographic and Geochemical Study of Coexisting Limestones and Dolostones from the Mascot-Jefferson City District, East Tennessee and the Right Fork Area, Central Tennessee," (Unpublished Master's Thesis, University of Tennessee, Knoxville).

Crawford, J. and Hoagland, A. D., 1968, "The Mascot-Jefferson City Zinc District, Tennessee," Ore Deposits of the United States, 1933-1967, J. D. Ridge, ed., AIME, New York, pp. 242-256.

Ebers, M. L., 1976, "A Study of Gangue Dolomite Mineralization in the Mascot-Jefferson City Zinc District, Using Cathodoluminescence," (Unpublished Master's Thesis, University of Tennessee, Knoxville, 82 pp.).

____________ and Kopp, O. C., 1979, "Cathodoluminescent Microstratigraphy in Gangue Dolomite, the Mascot-Jefferson City District, Tennessee," Economic Geology, Vol. 74, pp. 908-919.

Ferguson, T. L., 1981, "Petrographic and Trace-Element Characterization of Coarse Crystallization Carbonate Minerals in Fractures and Breccia Bodies and Associated Mineralization in the Right Fork Area, Central Tennessee," (Unpublished Master's Thesis, University of Tennessee, Knoxville, 81 pp.).

Graf, D. L., 1982, "Chemical Osmosis, Reverse Chemical Osmosis, and the Origin of Subsurface Brines," Geochimica et Cosmochimica Acta, Vol. 46, pp. 1431-1448.

Hoagland, A. D., 1976, "Appalachian Zinc-Lead Deposits," Handbook of Stata-Bound and Stratiform Ore Deposits, K. H. Wolfe, ed. Elsevier Science Publishers, Amsterdam, pp. 495-534.

______________, Hill, W. T. and Fulweiler, R. W., 1965, "Genesis of the Ordovician Zinc Deposits in East Tennessee," Economic Geology, Vol. 60, pp. 693-714.

McCormick, J. E. et al, 1971, "Environment of the Zinc Deposits of Mascot-Jefferson City District, Tennessee," Economic Geology, Vol. 66, pp. 757-762.

______________ et al, 1985, "Zinc Deposits of the Mascot-Jefferson City and Copper Ridge Districts, East Tennessee," Field Trips in the Southern Appalachians, 1985 SE GSA, University of Tennessee Studies in Geology, No. 9, p. 1-33.

Moss, T. A., 1982, "A Comparative Study of the Trace Element Distribution in Limestones, Dolostones, and Breccias in the Upper Knox Group of the Copper Ridge Zinc District, East Tennessee," (Unpublished Master's Thesis, University of Tennessee, Knoxville, 173 pp.).

Rittenhouse, G., 1967, "Bromine in Oil-Field Waters and Its Use in Determining Possibilities of Origin of These Waters," Bulletin American Association of Petroleum Geologists, Vol. 51, pp. 2430-2440.

Roedder, E., 1971, "Fluid Inclusion Evidence on the Environment of Formation of Mineral Deposits of the Southern Appalachian Valley," Economic Geology, Vol. 66, pp. 777-791.

Snyder, J. F., 1975, "Petrography and Minor Element Distribution in Wall Rocks from the Mascot-Jefferson City Zinc District," (Unpublished Ph.D. Dissertation, University of Tennessee, Knoxville, 209 pp.).

Viets, J. G., Mosier, E. L. and Erickson, M. S., 1983, "Geochemical Variations of Major, Minor, and Trace Elements in Samples of the Bonneterre Formation from Drill Holes Transecting the Virburnam Trend Pb-Zn District of Southeast Missouri," Proceedings, International Conference on Mississippi Valley-Type Lead-Zinc Deposits, University of Missouri, Rolla, pp. 174-186.

West, J. E., 1974, "Trace Element Geochemistry of the Coarse Crystalline Ore-Breccia Dolomite in the Flat Gap Mine, Treadway, Tennessee," (Unpublished Ph.D. Dissertation, University of Tennessee, Knoxville, 91 pp.).

Chapter 13

ORE MICROSCOPY AND GENESIS OF THE COPPER-SHALE DEPOSIT AT CRETA, OKLAHOMA

by Richard D. Hagni and Delbert E. Gann

Gulf Oil Foundation Professor of Geology, University of Missouri-Rolla
Rolla, Missouri

Associate Professor of Geology, Millsaps College
Jackson, Mississippi

ABSTRACT

Ore microscopic study of copper-bearing shale at the Creta mine in southwestern Oklahoma reveals a variety of copper sulfide grains and shows that each has originated by replacement of original constituents in the Permian Flowerpot Shale at Creta. The largest grains, about 120 μm in diameter, are replacements of megaspores with triletes morphology. Their interiors are shown by Auger microprobe analyses to be organic matter. The interior organic matter and the spore exines have been replaced by anilite and djurleite. Digenite and anilite replace collomorphic pyrite averaging 40 μm in diameter. Grains 10 μm or less in diameter are mostly euhedral, six-sided, platy crystals that have anilite compositions and are pseudomorphous after pyrrhotite. Other 10 μm copper sulfide grains are pseudomorphous after octahedra and cubes of pyrite. The results of this study indicate that the copper minerals are epigenetic and were deposited from fluids introduced into the Creta shale when it was relatively uncompacted and permeable. The shale served as a favorable reducing environment that contained a source of sulfur in the form of pyrite and pyrrhotite.

INTRODUCTION

The copper deposit at the Creta mine is the most important ore body of its type in Oklahoma, Texas, and Kansas. Although numerous occurrences of stratiform copper are known in the Permian sandstones and shales in those three states (Johnson, 1976a), the Creta deposit has been the most productive. Since its discovery in 1962 and the initiation of mining in 1965, the Creta mine has produced about 1.9 million short tons of ore at an average grade of 1.9% copper. Although the mine closed early in 1975, at one time it reached a production capacity of 1000 tons of ore per day (Dingess, 1976). The

Creta mine was the largest shale-hosted copper producer among the red bed copper deposits of the southwestern United States. Including the sandstone-hosted red-bed deposits in that portion of the country, Creta ranks third in total production behind that of the Stauber (5 million tons of ore) and Nacimiento (3.15 million tons of ore) mines in New Mexico (Jones, 1965).

The mine is located in southwestern Oklahoma near the railroad loading station of Creta about five miles north of the Texas state line and about 15 miles southwest of Altus, Oklahoma.

Although most of the Permian copper deposits of this region occur in red bed sandstone, the ore deposit at Creta is entirely contained within a greenish-gray shale of the Middle Permian Flowerpot Formation. In the Creta area, this formation is composed dominantly of red shale, but thin units of gypsum, dolomite, and gray to greenish-gray shale are distributed throughout the formation. All of the copper mineralization at Creta occurs in several thin layers that are contained within a greenish-gray shale that is five feet thick and located 30 to 35 feet below the top of the Flowerpot Formation. Examination of the outcrop shows no evidence to explain why this shale unit is mineralized and the other gray shales in the formation are not.

Ore-grade copper mineralization is restricted to two thin shale units within the five foot interval. The lowest of these is the main ore horizon, named the Prewitt Copper Shale, and it is eight to eight and one-half inches thick. This very thin ore horizon is conformable to and contained within two, thin, fine-grained, clastic units beneath a thin gypsum bed (Fig. 1). The lower host rock unit is a light bluish- to greenish-gray non-laminated or blocky mudstone that is 12 to 20 inches thick. The overlying host rock is a medium gray, laminated, silty shale, which is six to seven inches thick, and that pinches out westward in a lateral facies with bedded gypsum (Dingess,

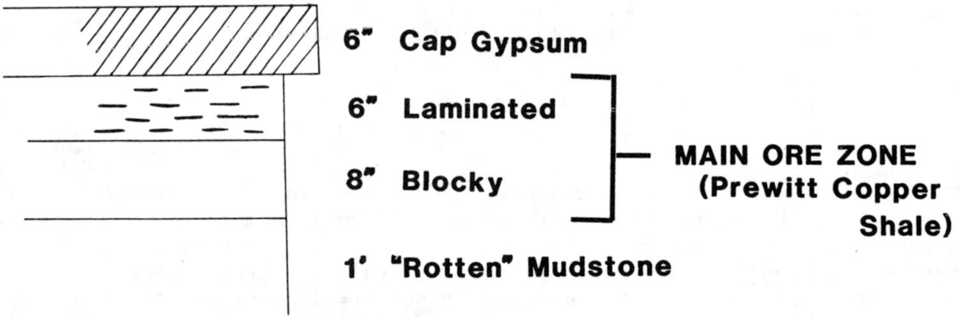

Fig. 1. Prewitt copper shale, Creta mine, southwestern Oklahoma.

1976). The richest copper ore, up to 4.5%, occurs in the upper four to five inches of the blocky mudstone and in the lower portion of the laminated shale. A bedded gypsum, about six inches thick and locally termed the "cap gypsum", overlies the laminated shale. The lower ore horizon has been named the Prewitt copper shale by Johnson (1976b).

A repetition of the above stratigraphic succession forms the host rock for the upper ore horizon located about five feet higher. The upper ore horizon is one to four inches thick, and only locally is its copper grade sufficient to mine.

Although the copper mineralization is restricted in its stratigraphic distribution, the lateral extent of the Prewitt copper shale is remarkable. Copper grades exceeding 0.5% are known to extend over a 15 square mile area about Creta (Dingess, 1976). Anomalous amounts of copper (500 to 1000 ppm) are present in this horizon of the Flowerpot for much greater distances along the strike of its outcrop to the north and south of Creta. Within the regional paleogeographic setting, Creta is located in a narrow northeast trending belt of clastic sediments between a landmass to the southeast and a Permian evaporite basin to the northwest (Johnson, 1976b).

The present investigation was undertaken to determine those features of the host rock that made it favorable for copper mineralization. The very fine-grained character of the ore necessitated the use of ore microscopic and petrographic techniques for the study, which showed that each of the various microscopic forms of copper grains present in the Creta ore is characterized by its own mode of formation and that knowledge regarding these forms is helpful in obtaining a better understanding of the genesis of these copper-shale deposits.

SPORE REPLACEMENTS

The spore replacement character of the ore, which has been described elsewhere (Hagni and Gann, 1976), is briefly reviewed here to outline its general aspect and to provide some new information. About one-third of the copper present in the Creta ore occurs in the form of replacements of megaspores of trilete morphology. Within the main ore zone almost all of these spores are completely replaced by copper sulfide, and there is no clear evidence that they originally were megaspores except for their size and shape. The replacements are mostly in the size range of 100 to 200 μm, and they are characterized by very smooth convex exterior margins. Their ellipsoidal shapes are especially suggestive.

The unmineralized portions of the blocky mudstone contain unreplaced megaspores, which in thin sections can be seen to contain an orange material. This material has previously been referred to as "body material" (Wilson, 1962; Hagni and Gann, 1976). Its orange color contrasts strongly with the light yellow color of the spore case or exine in thin sections. The orange material appears to be isotropic and does not yield an X-ray pattern. Quantitative reflectance measurements in polished sections yield $R_{546} = 2.3\%$, but the material does not take a good polish, and this figure may be lower than its actual reflectance. Its indentation hardness is $VHN_{50} = 30$ to 46 kg/mm^2. Scanning electron microscope energy dispersive spectroscopic (SEM-EDS) analyses of the orange material show the presence of a large amount of barium, significant sulfur, and smaller amounts of silicon and aluminum (Fig. 2a). Its composition differs from that of the spore case, which contains chlorine, silicon, and potasium (Fig. 2b). Based upon these SEM-EDS results, it is uncertain whether the orange material is an amorphous barite filling material or rather an organic material containing some barium and sulfur.

Subsequent analyses of spores mounted by pressing them into an indium foil by the Auger microprobe has conclusively shown that the material is composed dominantly of carbon, together with some oxygen and smaller amounts of barium and sulfur (Fig. 3a). Measurement of the Auger microprobe peaks divided by the sensitivity factors for each element gives semi-quantitative figures that show the orange material to consist of about 75 at. % carbon, 7.5% oxygen, 14% barium, 3.6% sulfur, and 0.1% potassium. Although these results are very approximate, the value for barium compares favorably with that of Dingess (1975) who obtained 15% barium in a chemical analysis of a sample of similar orange material that occurred in a vein at Creta. However, the amounts of all of these elements are very much on the high side due to the probable presence of hydrogen, an element that is not detected by the Auger microprobe. The indium peak is a contaminant resulting from its use as a mounting media. The shape of the oxygen peak is unusual and probably indicates that a significant portion of the oxygen is associated with the surface of the indium foil rather than with the carbon. Auger microprobe analyses of the spore cases show that they consist approximately of 65 at. % carbon, 23% oxygen, 4% calcium, 4% scandium, 2.5% sodium, 1.7% chlorine (Fig. 3b). The amount of hydrogen presumably present would reduce the percentages for all of the elements.

Eleven electron microprobe analyses (Table 1) of the copper sulfide that replaces spores indicate that its composition is similar to that of anilite ($Cu_{1.75}S$) rather than one of the other chalcocite group minerals: chalcocite (Cu_2S), djurleite ($Cu_{1.96}S$), and digenite (averages $Cu_{1.72}Fe_{0.025}S$). The average composition for 11 analyses shown in Table I is $Cu_{1.77}S$, a composition that is similar to that of anilite, $Cu_{1.75}S$. The analyses show a significant range in composition from $Cu_{1.73}S$ to $Cu_{1.82}S$. These results suggest that the copper sulfide is a fine-grained mixture of anilite and small, variable amounts of djurleite that have transformed into a relatively

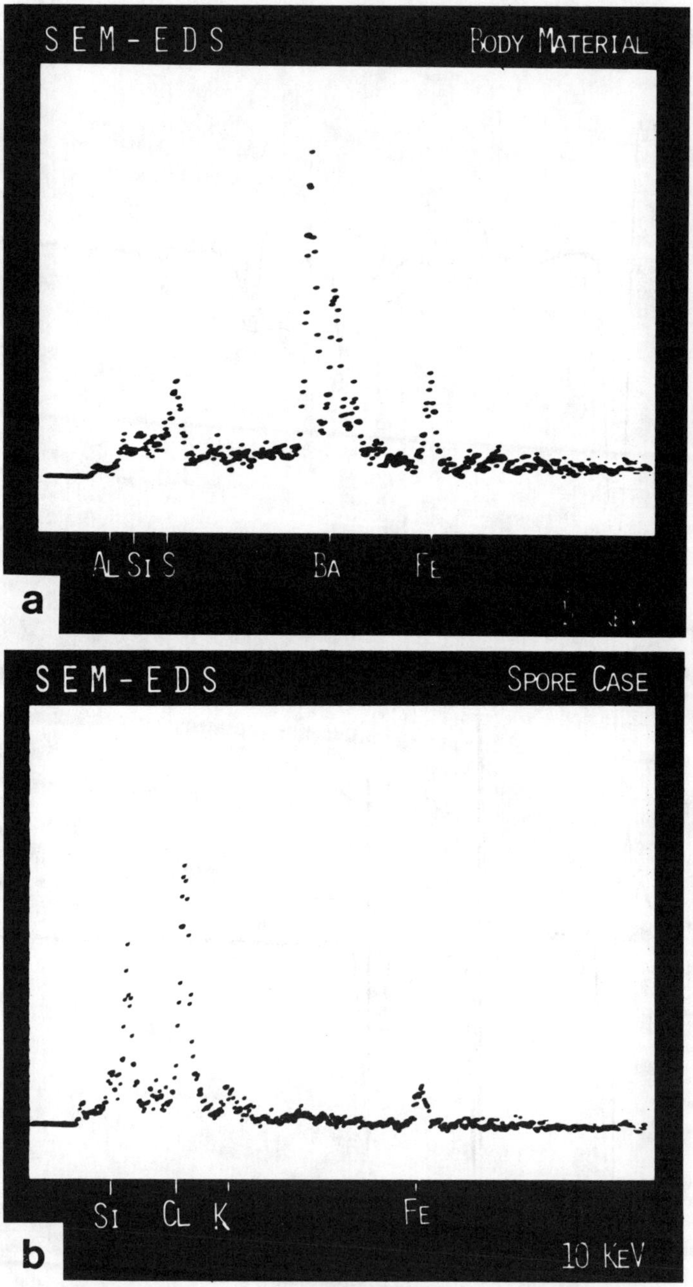

Fig. 2. Scanning electron microscope - energy dispersive analyses of the orange organic material (a) and spore case (b).

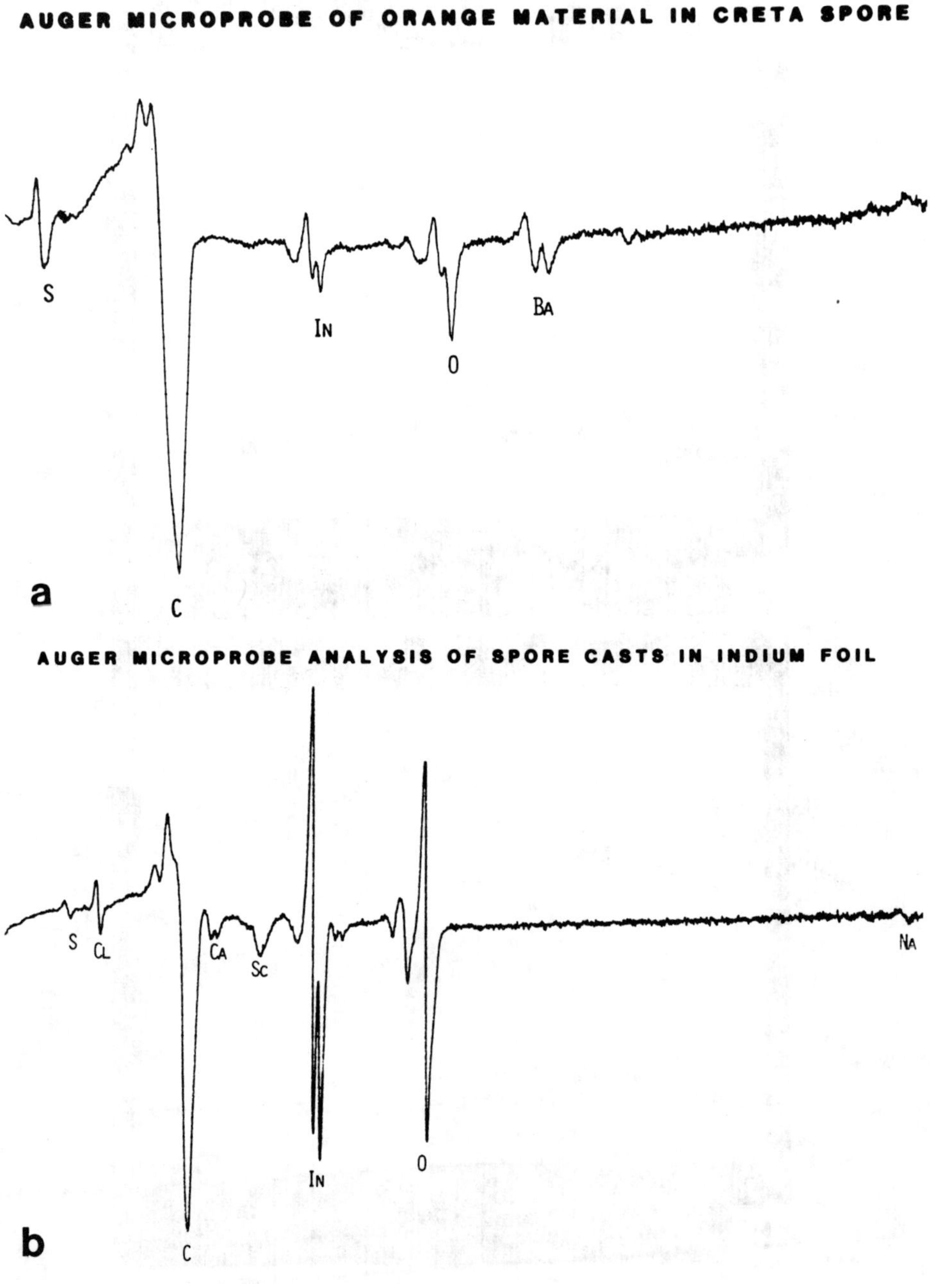

Fig. 3. Auger microprobe analyses of the orange organic material (a) and the spore case (b).

TABLE 1. Electron Microprobe Analyses of Copper Sulfide Minerals from the Creta Mine, Southwestern Oklahoma

Sample no.	Wt % Cu	Wt % Fe	Wt % Ag	Wt % S	Total	Formula	Notes
Anilite Replacements of Spores:							
1-6	77.4			21.41	98.8	$Cu_{1.82}S$	
1-18	76.5			21.47	98.0	$Cu_{1.80}S$	
5-15	77.1			21.74	98.9	$Cu_{1.79}S$	center
5-15	77.5			21.81	99.3	$Cu_{1.79}S$	edge
1-4	78.6			22.25	100.8	$Cu_{1.78}S$	
1-8	78.4			22.31	100.7	$Cu_{1.77}S$	
5-17	78.6			22.55	101.2	$Cu_{1.76}S$	
1-3	79.0			22.57	101.5	$Cu_{1.76}S$	case
1-3	78.7			22.64	101.3	$Cu_{1.75}S$	case
1-3	76.8			22.36	99.1	$Cu_{1.73}S$	dark
1-14	76.2			22.19	98.4	$Cu_{1.73}S$	
				Average of 11 grains =		$Cu_{1.77}S$	
Digenite Replacement of Pyrite:							
2-4	76.1	0.460		21.67	98.2	$Cu_{1.77}Fe_{0.012}S$	
5-28	75.4	0.970		21.60	98.0	$Cu_{1.76}Fe_{0.026}S$	
5-28	75.3	1.110		21.80	98.1	$Cu_{1.74}Fe_{0.029}S$	
5-8	76.1	0.393		22.03	98.5	$Cu_{1.74}Fe_{0.010}S$	
5-26	75.8	0.383		21.99	98.1	$Cu_{1.74}Fe_{0.010}S$	
2-4	77.9	0.173		22.69	100.7	$Cu_{1.73}Fe_{0.004}S$	
5-28	76.4	0.491		22.40	99.3	$Cu_{1.72}Fe_{0.013}S$	
				Average of 7 grains =		$Cu_{1.74}Fe_{0.015}S$	
Anilite Intergrown with Digenite Replacements of Pyrite:							
5-21	78.1	0.108		21.47	99.6	$Cu_{1.83}S$	
5-14	77.1	0.049		21.89	99.0	$Cu_{1.78}S$	
5-14	77.4	0.087		22.03	99.5	$Cu_{1.77}S$	
5-14	77.6	0.064		22.27	99.9	$Cu_{1.76}S$	
5-21	77.2	0.166		22.39	99.7	$Cu_{1.74}S$	
				Average of 5 grains =		$Cu_{1.78}S$	
Large Anilite Crystals:							
5-2	78.1			21.97	101.1	$Cu_{1.79}S$	
2-23	76.2			22.05	98.2	$Cu_{1.74}S$	
				Average of 2 grains =		$Cu_{1.77}S$	

uniform synthetic compound during polishing as has been found to be the case for other copper ores and some synthetic copper sulfide products (Morimoto et al., 1969). Ripley et al. (1980) have found that anilite is an important copper sulfide phase in the subeconomic copper occurrences in the stratigraphically lower, Permian Wellington Formation in south-central Kansas.

It is the organic material that is so favorable to copper sulfide replacement at Creta. At the base of the copper ore zone, all stages of replacement of the megaspores by copper sulfide can be observed (Figs. 4 and 5). The replacement process commonly begins in the central portions of the orange organic material and it progresses outward until all of the orange material and the spore case or exine is replaced by copper sulfide as well. The orange body material is preferentially replaced prior to the replacement of the spore case (Fig. 4b,c,d,e). Within the main ore zone the megaspores are completely replaced by copper sulfide (Fig. 4f). For other spores, the replacement is less regular and may begin in almost any part of the spore and in those instances of incomplete replacement almost any portion of the spore may remain as replacement remnants (Fig. 5b,c,d).

Although pyrite grains that occur in Permian rocks in Kansas have been interpreted by Lambert (1979), Lambert et al. (1980), and Ripley et al. (1980) to be megaspore replacements and to be similar to the sulfide replacements at Creta, most of those grains probably are not megaspore replacements. The copper sulfide grains in the Kansas occurrence do not have the characteristics of the Creta spore replacements, but rather they have a size of about 40 μm and a lumpy exterior morphology, features that are characteristic of the pyrite grains and their copper sulfide replacements at Creta, which are described in the following section. Progressive degrees of megaspore replacement apparently have not been observed in the Kansas occurrence. Where spores are clearly replaced by pyrite, the latter does not follow the outlines of the former (Ripley, 1984). At Creta, the replacement of spores by pyrite, rather than by copper sulfides, is a very rare occurrence.

Although most spore replacements appear under reflected light to be homogeneous, some have small portions that are brighter and less blue than the major portions. The lighter areas have slightly higher copper-sulfur ratios, but are not equivalent to that of chalcocite, Cu_2S. The copper sulfides that replace the spores can be distinguished from those which replace pyrite by the virtual absence of iron in the former grains. Microprobe analyses show that the grains of anilite and djurleite mixture are virtually lacking in iron content.

The copper sulfides that have replaced the spore cases usually are indistinguishable from those that have replaced the interior organic material in the megaspores. Some of the replaced exines, however, can be distinguished in SEM back-scatter images such as shown in Fig. 4f,

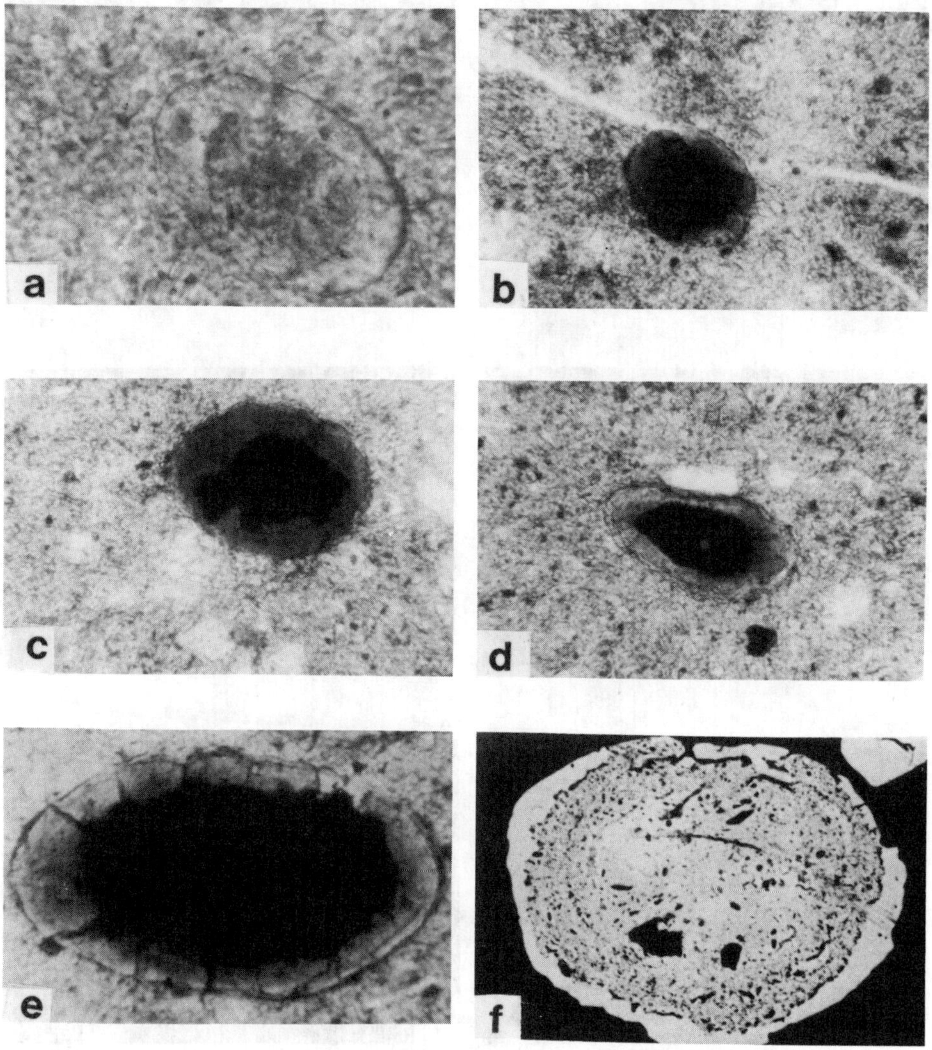

Fig. 4. Transmitted light photomicrographs showing progressive replacement of megaspores by anilite and djurleite. a - unreplaced spore; b,c,d - increasing amounts of the orange body material (medium gray) replaced by anilite (black); e - body material completely replaced and initial replacement of spore case; f - SEM photomicrography of completely replaced spore. All of the spores are about 120 μm long.

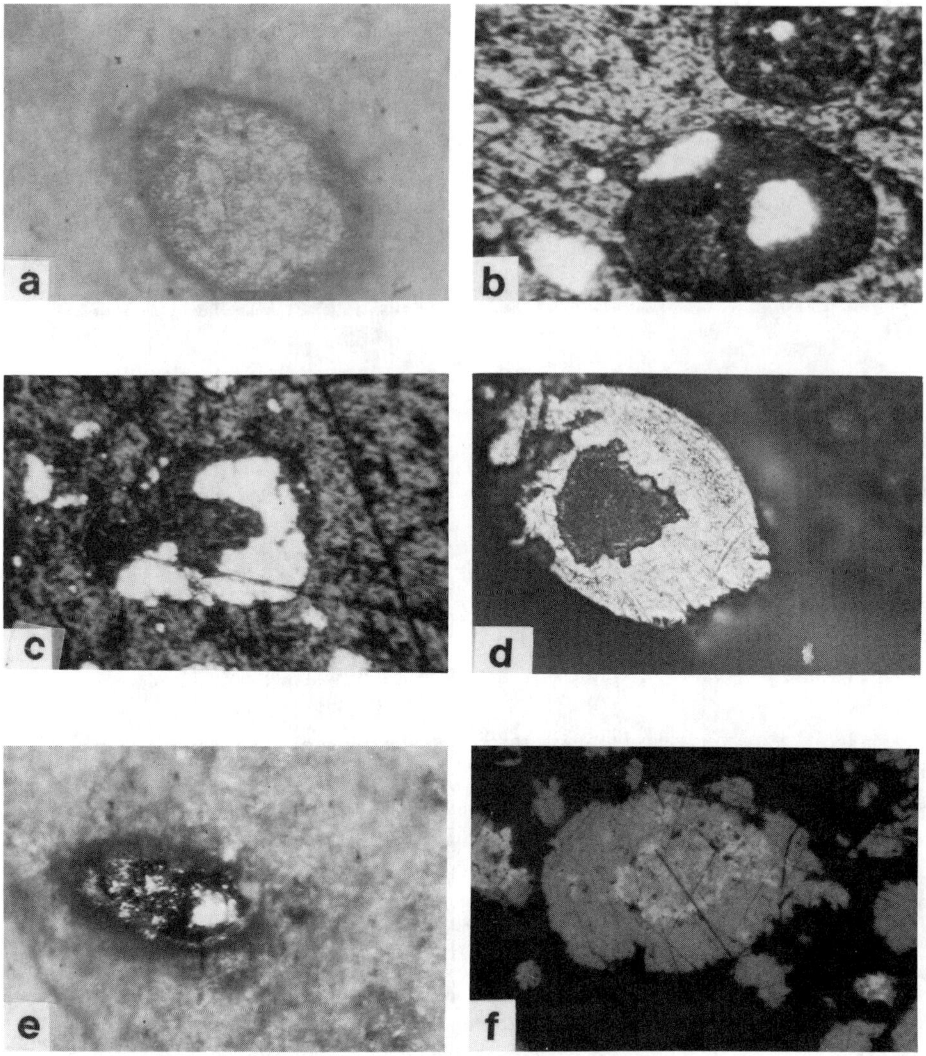

Fig. 5. Reflected light photomicrographs showing various degrees of replacement of megaspores by anilite and djurleite. a - unreplaced spore; b,c,d - partial replacement; e - ultrapac. objective view of anilite (black) within spore case; f - rare instance of spore replaced first by pyrite (white) and then by copper sulfide (grey). All of the spores are about 120 μm long.

or by their slightly greater hardness and tendency to develop oxidation skins. These properties suggest that the copper sulfide mixtures that replace some spore cases may contain somewhat greater amounts of djurleite.

The Creta trilete spores are characterized to a large extent by their being filled with the organic material and by a small degree of collapse. The spores also are remarkable in their lack of compaction. Most spores exhibit little or no compaction. This is surprising in view of the fact that the Blaine and other Formations, with a composite thickness of about 1,200 to 2,200 feet, formerly was present over the Flowerpot Formation at Creta and should have provided a sufficient load pressure to have strongly compacted the spores. The significance of these observations is that the organic matter prevented the spores from collapsing. Consequently, the organic material must have been present early in the diagenesis of the host rock, and it probably was an original constituent of the spores.

An important feature of the spores is their stratigraphic distribution. Although the copper mineralization diminishes downward from the main ore zone, the trilete megaspores, filled with the organic material, persist. This indicates that the presence of favorable spores and their contained organic material are not sufficient by themselves to form the copper ores.

PYRITE REPLACEMENTS

Since the presentation of our paper in 1976 at the symposium on Stratiform copper deposits of the midcontinent region, continued ore microscopic examination of Creta ores has contributed to a much better understanding of the iron sulfide replacements. The replacement of the iron sulfide is now considered to be even more important than originally realized, and the character of the unreplaced pyrite grains has been observed directly. Pyrite replacements are the most important form of copper sulfide grains in the Creta ores and comprise 50% or more of the ore.

The most abundant iron sulfide replacements are in the form of grains that are somewhat spherical in shape and 20 to 40 μm in size (Figs. 6 and 7). In the main ore zone, they are almost entirely replaced by copper sulfides (Fig. 6f) and can be recognized as iron sulfide replacements only by the presence of very tiny remnants of iron sulfide in the form of irregularly shaped areas and atolls (Fig. 6e). The iron sulfide replacements can be distinguished from spore replacements by their characteristic sizes and especially by their shapes. Unlike the spore replacements, their margins are rarely smooth. Instead, they are characterized by a lumpy or scalloped exterior surface with local V-shaped re-entrants.

The original, 20 to 40 μm, iron sulfide grains prior to copper

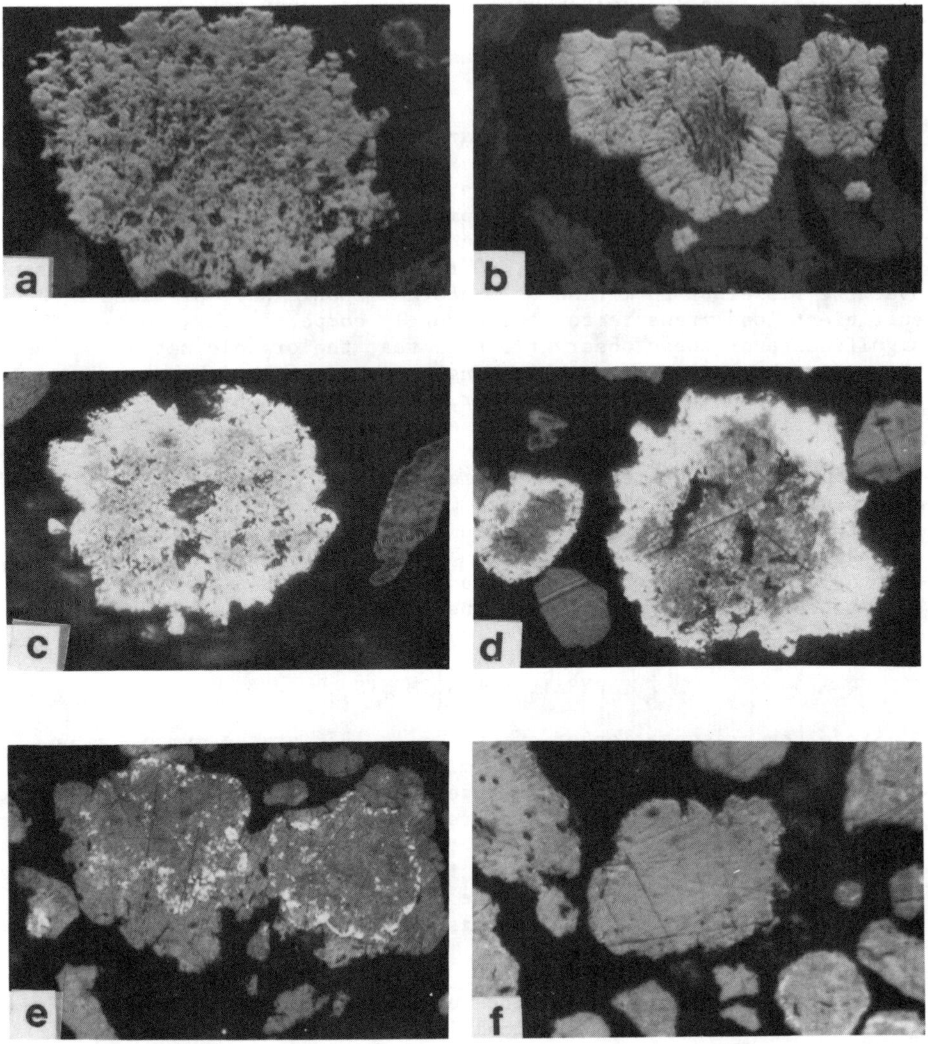

Fig. 6. Reflected light photomicrographs showing progressive replacement of colloform pyrite (white) grains by digenite (grey). a - virtually unreplaced pyrite; b,c,d - increasing amounts of replacement by digenite which takes place centrally to leave pyrite atolls; e - replacement remnants of pyrite arranged in an atoll; f - complete replacement with digenite showing the exterior morphology of the replaced pyrite grain. All pyrite and digenite grains are about 40 μm across.

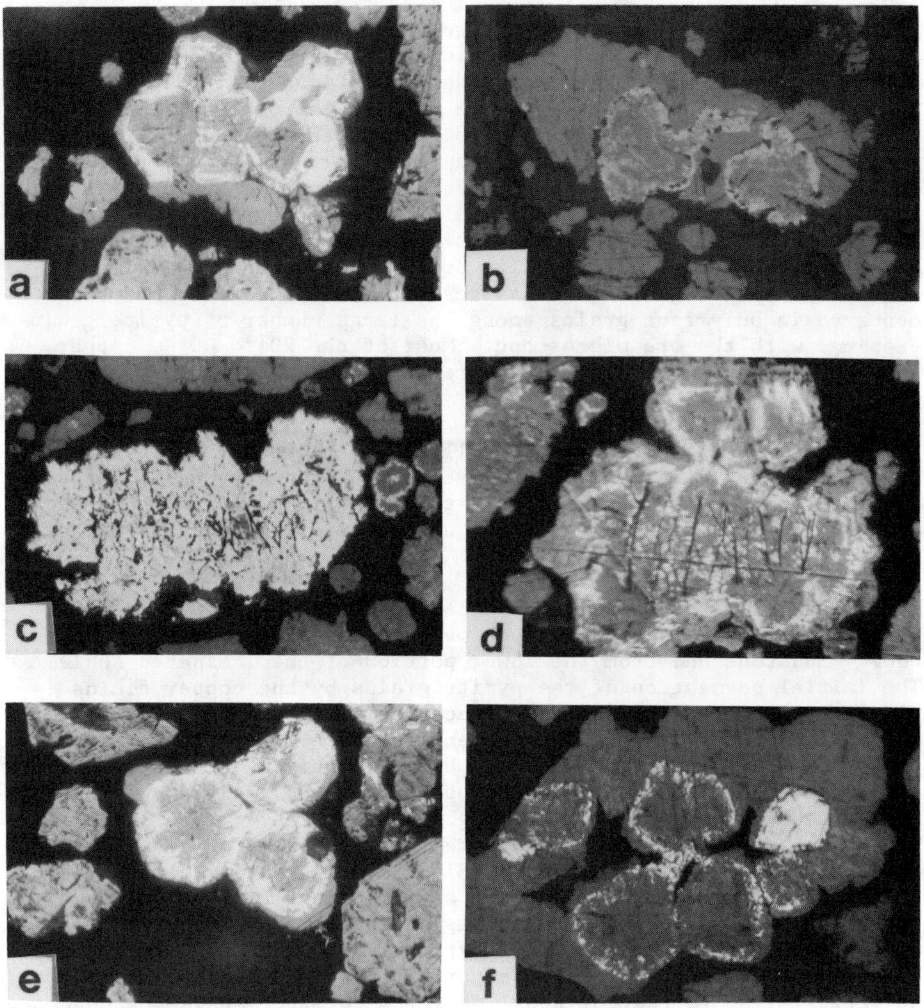

Fig. 7. Reflected light photomicrographs showing various degrees of replacement of pyrite (white) grains by digenite (grey). a - atolls of more crystalline pyrite present as replacement remnants in digenite; b - colloform pyrite partially replaced and overgrown by crystalline copper sulfide; c - colloform pyrite showing shrinkage cracks; d - digenite that has partially replaced colloform pyrite and retained the shrinkage cracks; e - cluster of three colloform pyrite grains partially replaced by digenite; f - atoll-shaped replacement remnants of a cluster of three former colloform pyrite grains. The pyrite grains are 20 to 60 μm across.

sulfide replacement consisted entirely or almost entirely of pyrite (Fig. 6a,b,c). Although all of this pyrite is now fine-grained, most of the grains consist of a central area of finer grained pyrite with successively coarser grained pyrite forming the outer portion (Fig. 6a,b,c,d). Some grains exhibit two or even three bands of repeated fine to coarse-grained iron sulfide (Fig. 7b). The contacts between the successive bands are characterized by a scalloped surface similar to that of their exterior shape. All of these properties are characteristic of colloform or melnicovite-pyrite. Although these grains have the same average size as those which are typical of framboidal pyrite grains, the typical framboidal texture has been observed in only four grains among the large number of pyrite grains examined with the ore microscope. Most of the 20 to 40 μm copper sulfide grains in the Creta ore were formed by the replacement of diagenetic, colloform pyrite.

The outer, more coarsely crystallized portions of some pyrite grains exhibit crude pyritohedral shapes. Although these forms have resulted from crystalline pyrite overgrowths in some instances, more commonly they simply represent the character of the outer, coarser colloform pyrite grains.

All stages of progressive replacement of the colloform pyrite grains have been observed in specimens collected from the top of the blocky mudstone and from the lower portion of the laminated shale. The initial permeation of the pyrite grains by the copper fluids is along tiny shrinkage crack-like fractures that form an irregular reticulate pattern throughout the entire grain. The fine-grained pyrite of the interior portion had been selectively replaced over that of the coarser grained outer portions during the early stages of replacement by copper sulfides (Fig. 6b,c,d). As replacement progressed, the coarser grained outer portions were gradually reduced to the form of continuous atolls (Fig. 6d; Fig. 7a,b,d,e) and then to rings of "sea islands" in an atoll distribution (Fig. 6e; Fig. 7f). This was succeeded by an occasional irregularly shaped pyrite remnant or two, and finally the original pyrite grain became entirely converted to copper sulfide. The origin of Creta atoll textures may be very similar to that of atoll textures in other shale-hosted copper deposits, such as those described by Schouten (1946) at Mt. Isa, Australia.

Continued ore microscopic study of the colloform pyrite grains has shown that they occur in a wide range of sizes and that some are partly clustered together to form nearly solid pyrite masses. Although the most common pyrite replacements are 20 to 40 μm across, these grains exhibit a size range that extends down to the smallest observable with the ore microscope, only a few micrometers across. The colloform pyrite grains of all sizes, but especially the smaller ones, exhibit a tendency to be clustered into groups of two or more grains (Fig. 7e,f). These clusters of colloform pyrite are not only replaced but they may also serve as the nucleus for the overgrowth of larger copper sulfide crystals (Fig. 7b).

Some of the 20 to 40 μm pyrite grains are better crystallized, especially in their outer portions where they exhibit crystal faces of the pyritohedron. Copper sulfide replacements of these grains typically exhibit six sides and some contain atolls of pyrite with similar shapes (Fig. 7a). Rarely well crystallized pyrite was present and forms replacement remnants in the copper sulfide grains of this size (Fig. 7f).

Analyses by SEM-EDS of the pyrite grains has detected no trace elements other than the major constituents, iron and sulfur. Thus, the Creta pyrite grains appear to differ in this respect from colloform pyrite grains in some occurrences that contain significant amounts of cobalt, nickel, manganese, zinc, arsenic, and other elements.

Copper replacement of pyrite diminishes upward above the main ore zone at Creta, even though the pyrite persists. This direction is opposite to that of the spore replacements, which diminish downward. The authors' interpretation of this relationship is that both organic matter-filled trilete spores and the fine-grained pyrite were required to be present in the host shale in juxtaposition in order for the copper sulfides to be deposited. More than enough sulfur for the copper sulfide replacements of the pyrite was already present in the iron sulfide grains. The excess sulfur released by the replacement was available to combine with the copper to form the replacements in the nearby organic matter. These considerations also indicate that the wide range of sulfur isotope ratios determined for the Creta ores (Lockwood, 1976) are characteristic of the genesis of the sulfur obtained from the replaced iron sulfides and are not indicative of the genesis of the copper.

The principal copper sulfide phase present as pyrite replacements is one which is darker and distinctly more blue than the anilite and djurleite replacements of the megaspores and their contained organic matter. Only digenite among the chalcocite group minerals is darker and bluer than anilite. Electron microprobe analyses confirm that this phase is digenite. Its copper-sulfur ratios range from 1.72 to 1.77 and average 1.74 (Table I). Small amounts of iron, averaging 0.015 and up to 0.029 at. %, are present. Six analyses for this variety of Creta digenite are given in Table 1. Small amounts of iron (averaging 0.025 at. %) have been found to be characteristic of synthetic and natural digenite (Morimoto et al., 1969). In this respect, the digenite that replaces pyrite grains differs from the iron-free anilite that replaces spores.

Small patches of a brighter chalcocite group mineral are intergrown with the blue digenite in some pyrite replacement grains. Electron microprobe analyses indicate that it is anilite. Its average copper-sulfur ratio, 1.78, is higher than that of the digenite and similar to that of the anilite (1.77) that replaces spores, and it is iron-free.

In view of the abundant amount of pyrite in the Prewitt shale, it might be anticipated that pyrite would also have commonly formed replacements of the trilete megaspores, but pyrite replacements of the spores are almost entirely lacking. This indicates that the contained organic matter had not yet reached a state which would promote the deposition of sulfide at the time of formation of diagenetic pyrite. A few spores have been replaced by colloform pyrite prior to the deposition of the bluish copper sulfide, as shown in Fig. 5f. Such oocurrences indicate that the organic matter was present at the time of pyrite formation.

COPPER SULFIDE CRYSTALS

One of the major modes of occurrence of copper sulfide in the Creta ore is in the form of tiny crystals most of which are less than 10 μm in their longest dimension (Fig. 8a,b,c,d,e). Approximately 20% of the ore is comprised of these small crystals. Their extremely small size makes their study difficult, and a variety of techniques have been utilized. Although their shapes can be discerned with difficulty with the aid of an ore microscope, they are best studied by separation from the shale and observation with an ultrapac objective (Fig. 8c) or the scanning electron microscope (SEM) (Fig. 8d,e).

The most common shape is that of a six-sided platy crystal (Fig. 8a,d,e). The basal pinacoid is the best developed face, but two or three sets of unequally developed bipyramidal faces and a prismatic face are commonly present (Fig. 8b,c). These crystals probably are monoclinic and pseudo-hexagonal. Although three of the chalcocite group minerals, chalcocite, djurleite, and anilite, will form tabular crystals, the Creta crystals lack the laminated morphology shown in the SEM photomicrographs of typical djurleite from Calabona, Sardinia (Bernardini et al., 1973). Furthermore, electron microprobe analyses eliminate chalcocite and pure djurleite from further consideration. Thirteen analyses of these tabular crystals from the Creta ores show that they have copper-sulfur ratios that range from 1.71 to 1.83 and have an average composition of $Cu_{1.78}S$ (Table 2). These ratios and the lack of iron suggest that the crystals were composed mainly of anilite with small amounts of djurleite prior to polishing, similar to that found for the copper sulfide replacements of megaspores. Measurements of the interfacial angles in photomicrographs looking along the c-axis gives values close to 120 and 60 degrees. The angles indicate that the Creta crystals are pseudohexagonal and probably monoclinic, rather than pseudotetragonal and orthorhombic as reported for anilite (Morimoto et al., 1969). These results suggest that the crystal forms of the copper sulfide at Creta are pseudomorphous after some other mineral.

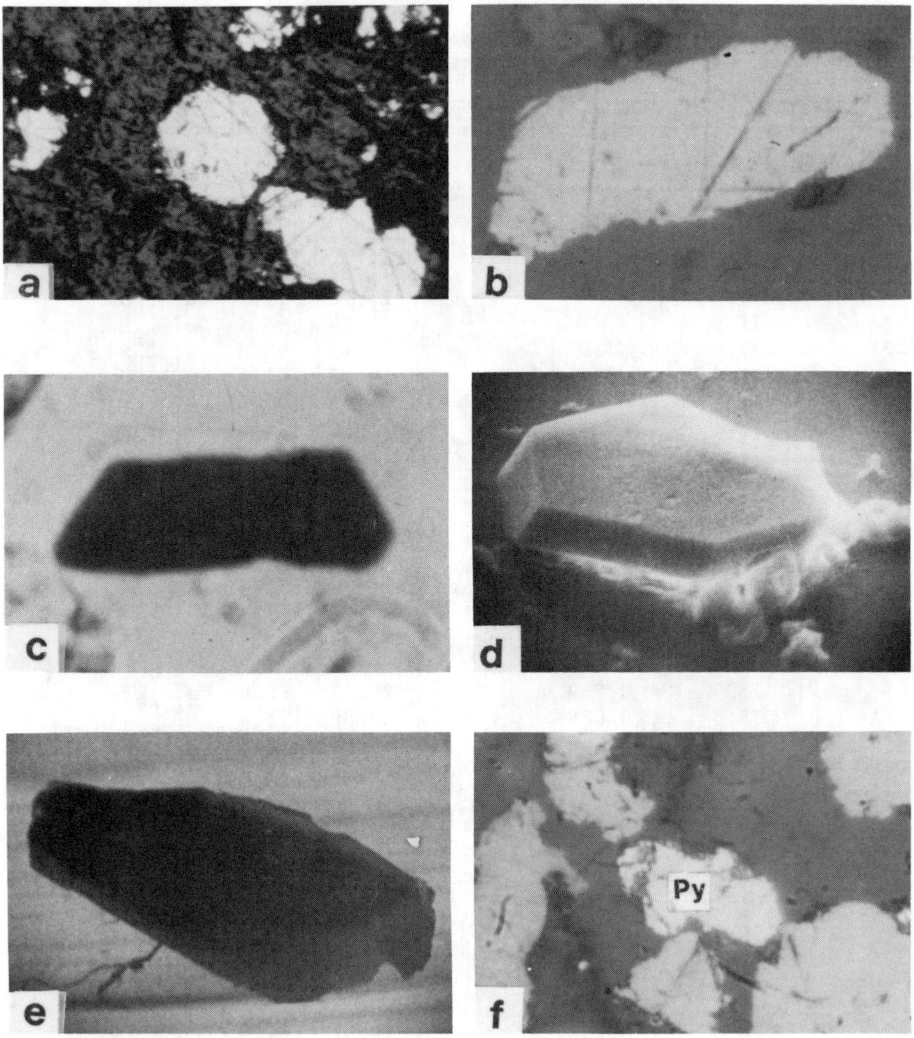

Fig. 8. Photomicrographs of 10 μm tabular copper sulfide crystals. a,b - reflected light; a - view showing six-sided character; b - view showing platy character; c - crystal on glass plate showing typical terminations on tabular crystal; d,e - SEM photomicrographs; f - tabular pyrite (Py) crystal from Creta ores that shows a similar morphology.

TABLE 2. Electron Microprobe Analyses of Copper Sulfide Minerals from the Creta Mine, Southwestern Oklahoma

Sample no.	Wt % Cu	Wt % Fe	Wt % Ag	Wt % S	Total	Formula
Six-sided, Tabular, 10 μm Copper Sulfide (Anilite) Grains:						
2-18	79.2			21.76	101.0	$Cu_{1.83}S$
2-25	78.5			21.91	100.5	$Cu_{1.81}S$
2-9B	77.8			21.82	99.6	$Cu_{1.80}S$
2-9	77.4			21.76	99.1	$Cu_{1.79}S$
2-26	77.2			21.74	98.9	$Cu_{1.79}S$
2-21	77.0			21.75	98.8	$Cu_{1.78}S$
2-13	78.6			22.36	101.0	$Cu_{1.77}S$
2-9A	78.5			22.40	100.6	$Cu_{1.76}S$
5-22	78.1			22.46	100.6	$Cu_{1.75}S$
2-11	76.9			22.29	99.2	$Cu_{1.74}S$
2-12	78.4			22.68	101.1	$Cu_{1.74}S$
2-21	77.3			22.38	99.9	$Cu_{1.74}S$
2-10	76.8			22.69	99.5	$Cu_{1.71}S$
				Average of 13 grains =		$Cu_{1.78}S$
Digenite Associated with Galena:						
5-3	75.3	0.569		22.77	98.6	$Cu_{1.67}Fe_{0.014}S$
1-7A	75.5	0.088		22.89	98.4	$Cu_{1.66}Fe_{0.002}S$
5-3	74.9	0.094		23.05	98.8	$Cu_{1.64}Fe_{0.002}S$
2-5	75.2	0.003		23.17	98.4	$Cu_{1.64}Fe_{0.001}S$
				Average of 4 grains =		$Cu_{1.65}Fe_{0.005}S$
Blaubleibender Covellite I Associated with Galena:						
1-7B	68.65			30.07	98.7	$Cu_{1.15}S$
Digenite Replacement of Bornite:						
2-24	78.3	0.232		22.51	101.0	$Cu_{1.75}Fe_{0.006}S$
2-24	77.5	0.075		22.28	99.8	$Cu_{1.75}Fe_{0.002}S$
1-12	75.7	0.885		21.92	98.4	$Cu_{1.74}Fe_{0.022}S$
1-12	76.1	0.784		22.10	98.9	$Cu_{1.74}Fe_{0.020}S$
2-1	76.0	0.85		22.16	98.9	$Cu_{1.73}Fe_{0.022}S$
				Average of 5 grains =		$Cu_{1.74}Fe_{0.014}S$

Since our earlier study (Hagni and Gann, 1976), it has been found that the tiny Creta crystals also have an important relationship to various forms of diagenetic iron sulfides. Ore microscopic study shows that they commonly contain pyrite replacement remnants, very similar in character to those in the 20 to 40 μm grains (e.g., Fig. 7b), but that they are smaller. In fact, there appears to be a continuation in grain size of those colloform bodies down to the smallest that can be detected with the microscope. In many grains, the colloform pyrite formed the site of initial copper sulfide replacement and subsequently became the nucleus for euhedral copper sulfide overgrowths.

Other tabular copper sulfide crystals originated through the pseudomorphic replacement of tabular iron sulfide crystals. A few unreplaced tabular pyrite crystals are present in Creta ore (Fig. 8f). These have unequal bipyramidal faces that are similar to those of the tiny copper sulfide crystals. Williams (1976) has observed similar tabular six-sided pyrite crystals in SEM photomicrographs of the Mt. Isa ore. Probably the tabular pyrite crystals themselves inherited their morphology from some previous iron sulfide phase that occurred early in the diagenesis of the host sediment. Two likely candidates are monoclinic pyrrhotite ($Fe_{0.87}S$) and mackinawite (FeS, tetragonal). These minerals form as intermediate products in the development of pyrite in sediments (Goldhaber and Kaplan, 1974) and can occur as tabular crystals. Euhedral pyrrhotite crystals with strikingly similar crystal shape and size are common in portions of the Rammelsberg, West Germany massive sulfide ore deposits. An average of 12 electron microprobe analyses for the Rammelsberg pyrrhotite gave a composition $Fe_{0.87}S$. This composition corresponds to monoclinic pyrrhotite (Arnold, 1969). Although no pyrrhotite has been detected in our ore microscopic study of the Creta ores, the very similar monoclinic pyrrhotite morphology, the fact that monoclinic pyrrhotite can crystallize as an intermediate iron sulfide product in sediments, and the presence of pyrite replacement remnants in the copper sulfide crystals suggests that the tabular 10 μm copper sulfide crystals have formed principally as pseudomorphic replacements of monoclinic crystalline pyrrhotite.

The second most abundant 10 μm copper sulfide grain in the Creta ore has the form of an octahedron. This type of grain was first recognized in an SEM study conducted by Kidwell and Bower (1976) on cores taken from drill holes south of Creta in Hardeman County, Texas. Some of the tiny octahedra in the Creta ore, when examined with the aid of the ore microscope, can be seen to have substantial amounts of unreplaced crystalline pyrite (Fig. 9a). Thus, this form has replaced diagenetic octahedral pyrite, or perhaps its metastable predecessor octahedral greigite. Less commonly, the copper sulfides form pseudomorphs after tiny cubic (Fig. 9b) and pyritohedral pyrite crystals. Framboidal pyrite, as shown in Fig. 9c, has rarely been found as replacement remnants in copper sulfide grains of this or any size in the Creta ores.

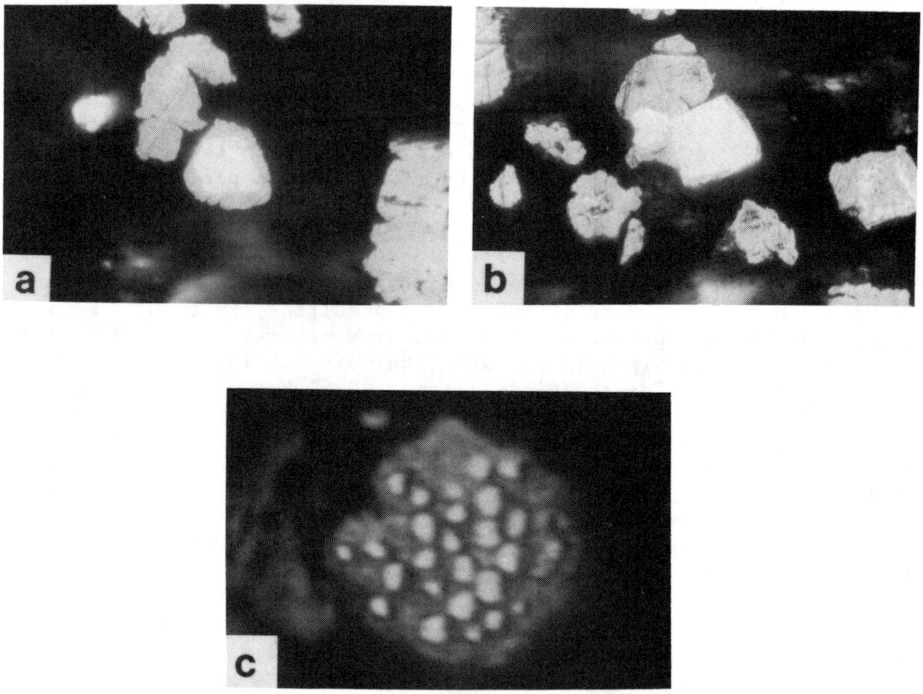

Fig. 9. Reflected light photomicrographs of digenite (grey) replacements of other 10 µm pyrite (white) forms. a - octahedral pyrite; b - cubic pyrite; c - framboidal pyrite. All of the digenite grains are about 10 µm across.

MINOR FORMS OF COPPER SULFIDE

Copper sulfide crystals that are larger than the spore and colloform pyrite replacements locally are present in the Creta ore. Electron microprobe analyses for two crystals indicate that they are anilite (Table 1). Their average copper-sulfur ratio of 1.77 is similar to that of the anilite that replaces spores, and the crystals are iron-free.

Minor veinlets of chalcocite locally have been observed in cracks, which exhibit a crude pattern that is similar to that of mud cracks (Johnson and Brockie, 1973). Veinlets are rarely observed in polished sections where they traverse and are parallel to the bedding of the host shale. Such veinlets attest to the migration of the copper-bearing fluids. The copper sulfide minerals also are locally concentrated in silty lamellae (Hagni and Gann, 1976).

ADDITIONAL SULFIDE MINERALS

In addition to the principal forms of copper, copper-iron, and iron sulfides discussed above, smaller amounts of 10 other sulfide minerals and digenite in two other occurrences are present. Eight of these have been identified subsequent to our earlier study (Hagni and Gann, 1976). The other minerals are galena, chalcopyrite, bornite, covellite, blaubleibender covellite I, stromeyerite, argentiferous djurleite, argentiferous chalcocite, carrollite, and native bismuth.

Galena

Galena is the most abundant of the eleven additional sulfide minerals in the shale samples studied to date. The Creta ore contains an average of 0.05 wt. % lead (Dingess, 1976). Galena occurs in grains that are separate from those of the other copper sulfide minerals except for covellite. Some of the smaller galena grains, those that are about 10-15 m across, exhibit six-sided shapes similar to those noted for the smallest copper sulfide grains that are interpreted to have formed by replacement of pyrrhotite. Galena commonly has formed as a coating and partial replacement of grains of digenite and blaubleibender covellite I (Fig. 10a). Thus, galena is a relatively late mineral in the Creta deposits where it was deposited after the iron sulfide and copper sulfide minerals.

Covellite

Covellite is about as abundant as galena in those samples that have been best studied. Most covellite occurs in grains that are separate from the other copper sulfide minerals. Covellite rarely contains remnants of small pyrite grains that are partially replaced. All of these covellite grains that were free from galena and that were examined under oil immersion showed the violet-red color of anisotropism for the omega direction that is typical for covellite, rather than the deep blue of blaubleibender or blue-remaining covellite. This feature, together with its occurrence in primary ores that lack other supergene minerals, suggests that the covellite is primary in origin.

Digenite Associated with Galena

A deep blue copper sulfide mineral is commonly associated with galena. Its typical textural relationship is one in which covellite has formed a core and galena formed the rim (Figure 10a). In those grains galena has formed later than covellite and has coated and partially replaced the earlier deposited copper sulfide. Electron microprobe analyses of these grains indicate that the blue phase is digenite. Four analyses show a narrow range of copper-sulfide ratios with an average of 1.65 and small amounts of iron (Table 2). The phase is bluer and has a lower copper-sulfur ratio than the digenite that replaces colloform pyrite.

Blaubleibender Covellite I

Some of the blue copper sulfide that forms replacement remnants in galena have a composition, $Cu_{1.15}S$ (Table 2), that is similar to blaubleibender covellite I, $Cu_{1.1}S$, of Potter (1977). The copper sulfide has the deep blue color of covellite but remains blue under oil immersion. Galena and the digenite and blaubleibender covellite that are associated with galena probably have formed at a relatively late time. The ore fluids may have been somewhat impoverished in copper and more oxidizing at that time.

Chalcopyrite

Chalcopyrite is an uncommon sulfide mineral in the Creta ores. Those grains that have been found commonly show exsolution intergrowth with bornite (Fig. 10b). Chalcopyrite may serve as the host for bornite or vice-versa. Some of those grains have formed as a replacement of trilete megaspores. Chalcopyrite in this occurrence has not been observed to form as a replacement of pyrite nor to contain pyrite inclusions. Four electron microprobe analyses of this chalcopyrite give an average composition of $Cu_{1.08}Fe_{0.99}S_2$ (Table 3).

Less commonly chalcopyrite replaces fractured pyrite (Fig. 10c). Chalcopyrite in this occurrence has a distinctly brown tint under oil immersion. An electron microprobe analysis indicates a slightly higher iron to copper ratio with a composition of $Cu_{0.97}Fe_{1.03}S_2$.

Bornite

Bornite also is an uncommon ore mineral in the Creta copper-shale Most of the bornite that has been recognized to occur in the Creta ore is present in exsolution intergrowth with chalcopyrite. It occurs as lamellae in grains of chalcopyrite, or as the host for exsolution chalcopyrite as shown in Fig. 10b. Part of the bornite and chalcopyrite in those occurrences has formed as replacements of trilete megaspores. Electron microprobe analyses of four bornite grains give an average composition of $Cu_{4.77}Fe_{0.93}S_4$ and indicate that

TABLE 3. Electron Microprobe Analyses of Copper-Iron Sulfide Minerals from the Creta Mine, Southwestern Oklahoma

Sample no.	Wt % Cu	Wt % Fe	Wt % Ag	Wt % S	Total	Formula
Bornite:						
1-12	60.23	26.78		11.49	98.5	$Cu_{4.53}Fe_{0.98}S_4$
2-1	63.33	25.88		10.52	99.7	$Cu_{4.93}Fe_{0.93}S_4$
2-1	64.17	26.09		9.55	99.8	$Cu_{5.28}Fe_{0.84}S_4$
			Average of 3 grains =			$Cu_{4.91}Fe_{0.92}S_4$
Chalcopyrite:						
1-12	35.35	29.41		34.67	99.4	$Cu_{1.03}Fe_{0.97}S_2$
1-12	35.86	28.86		34.04	98.7	$Cu_{1.06}Fe_{0.97}S_2$
1-12	36.60	28.54		33.33	98.4	$Cu_{1.11}Fe_{0.98}S$
1-12	35.71	29.16		33.71	98.5	$Cu_{1.07}Fe_{0.99}S$
2-1	35.50	29.12		34.37	98.9	$Cu_{1.04}Fe_{0.97}S_2$
2-1	38.13	26.95		32.83	97.9	$Cu_{1.17}Fe_{1.03}S_2$
			Average of 6 grains =			$Cu_{1.08}Fe_{0.99}S_2$

the bornite at Creta is a sulfur-rich variety. Three of those analyses are given in Table 3. The color in reflected light, however, is that of normal bornite rather than that of orange bornite.

Bornite also has been observed to form exsolution lamellae within argentiferous djurleite.

The smallest grains of bornite, those that are about 10-15 μm across, may contain very small replacement remnants of pyrite, showing that they have formed as pyrite replacements. Some of these grains, have a six-sided morphology similar to that shown by the 10 μm copper sulfide grains.

Digenite associated with Bornite and Chalcopyrite

In addition to the abundant digenite that has formed as a result of the replacement of the pyrite grains, very minor amounts of digenite occur as marginal replacements of bornite that is in exsolution intergrowth with chalcopyrite (Fig. 10b). Five analyses for digenite in those grains shows an average amount of 0.014 at. % iron and with a composition of $Cu_{1.74}Fe_{0.014}S$ (Table 2). Their average composition is very similar to that of the digenite that replaces pyrite.

Stromeyerite

Small amounts of silver were recovered at the smelter from the copper concentrates produced at Creta. The Creta copper ores have an average silver content of 0.15 oz. per ton and the copper concentrates ran 3-5 oz. silver per ton (Brockie, 1984). The silver-copper sulfide mineral, stromeyerite, reported here for the first time, is the principal silver-bearing mineral in the Creta ores. Stromeyerite is brighter and slightly softer than the associated copper sulfide minerals. Although it can easily be overlooked under the ore microscope, it can be recognized by its violet, pink, and tan reflective pleochroic color tints and by its strong anisotropism under oil immersion with purple, light blue, and tan colors of anisotropism. Stromeyerite is readily distinguished from the copper sulfide minerals in a SEM-back scatter image as shown in Fig. 10d.

Stromeyerite occurs as groups of intergrown small anhedral grains that are one to five μm across. It is closely associated with the bright, white copper sulfide minerals, argentiferous djurleite and argentiferous chalcocite. Some of the stromeyerite appears to have been replaced by djurleite and chalcocite as shown in Fig. 10d. The copper sulfides form a partial rim, veins, and contain tiny replacement remnants of the stromeyerite. Because the stromeyerite grains are contained in primary copper ores that exhibit no trace of supergene minerals, the stromeyerite that is present in the Creta ores appears to be hypogene in origin. Part of the stromeyerite and associated argentiferous djurleite occur as triletes megaspore replacements.

Electron microprobe analyses of the stromeyerite grains confirm that the mineral is a copper-silver sulfide. Analyses for five of the grains are given in Table 4. They show a composition similar to that previously reported for stromeyerite and unlike that of the optically similar copper-silver minerals, mckinstryite and jalpaite (Skinner, 1966).

Argentiferous Djurleite

The copper sulfide grains that are intergrown with stromeyerite are brighter and whiter than any other copper sulfide present in the Creta ores (Fig. 10d). Some of the electron microprobe analyses of those grains give an average copper-sulfur ratio of 1.94 and show that they contain small amounts of silver (Table 4). The copper-sulfur ratio determined for this phase is similar to that of djurleite, $Cu_{1.95}S$.

Between 0.5 and 1.7 wt. % silver was detected in the copper sulfide grains. The six analyses were determined on djurleite grains that appear to be free from inclusions of stromeyerite. The silver appears to occur in solid solution in the djurleite.

TABLE 4. Electron Microprobe Analyses of Silver-Copper Sulfide Minerals from the Creta Mine, Southwestern Oklahoma

Sample no.	Wt % Cu	Wt % Fe	Wt % Ag	Wt % S	Total	Formula
Stromeyerite:						
7-37	35.7		49.2	15.29	100.1	$Cu_{1.18}Ag_{0.96}S$
7-42	35.6		47.9	15.50	99.0	$Cu_{1.16}Ag_{0.92}S$
7-40	35.0		49.7	15.43	100.1	$Cu_{1.14}Ag_{0.96}S$
7-6	33.5		49.6	15.24	98.3	$Cu_{1.10}Ag_{0.97}S$
7-1	33.7		48.9	15.47	98.1	$Cu_{1.10}Ag_{0.94}S$
			Average of 5 grains =			$Cu_{1.14}Ag_{0.95}S$
Argentiferous Djurleite Associated with Stromeyerite:						
7-38	77.6		1.67	20.07	99.3	$Cu_{1.95}Ag_{0.03}S$
7-41	78.0		1.21	20.17	99.4	$Cu_{1.95}Ag_{0.02}S$
7-1	79.6		1.13	20.58	101.3	$Cu_{1.95}Ag_{0.02}S$
7-37	78.3		0.85	20.31	99.4	$Cu_{1.94}Ag_{0.01}S$
7-27	77.2		0.93	20.26	98.3	$Cu_{1.92}Ag_{0.01}S$
7-19	77.4		0.52	20.36	98.2	$Cu_{1.92}Ag_{0.01}S$
			Average of 6 grains =			$Cu_{1.94}Ag_{0.017}S$
Argentiferous Chalcocite Associated with Stromeyerite:						
7-17B	79.2		1.64	20.14	100.9	$Cu_{1.98}Ag_{0.02}S$
7-17C	78.8		1.57	20.00	100.4	$Cu_{1.99}Ag_{0.02}S$
7-23	80.1		0.94	20.04	101.1	$Cu_{2.01}Ag_{0.01}S$
			Average of 3 grains =			$Cu_{1.99}Ag_{0.014}S$

Argentiferous Chalcocite

Part of the copper sulfide grains that are intergrown with stromeyerite gave compositions from electron microprobe analyses that have an average copper-sulfur ratio of 1.99 (Table 4), a ratio close to that of chalcocite, Cu_2S. Three grains, which were free from stromeyerite, contained between 0.9 and 1.6 wt. % silver.

No other silver minerals have been found in the Creta ores, and SEM-EDS analyses have failed to detect any silver in solid solution in the other copper sulfide minerals that are not associated with stromeyerite nor in pyrite.

Carrollite

Rare grains of the cobalt-copper thiospinel, carrollite, have recently been detected in the Creta ores as a result of continued ore microscopic study. Carrollite is present as very small grains, about five micrometers across, which occur within grains of chalcopyrite adjacent to bornite exsolution lamellae (Fig. 10b). Small amounts of nickel and iron also are present in the thiospinel. Three electron microprobe analyses (Table 5) of one carrollite grain gave an average composition of $(Co_{1.88}Cu_{1.19}Fe_{0.10}Ni_{0.11})S_4$.

TABLE 5. Electron Microprobe Analyses of Thiospinel Mineral from the Creta Mine, Southwestern Oklahoma

No	Wt % Co	Wt % Cu	Wt % Fe	Wt % Ni	Wt % S	Total	Formula
Carrollite:							
1	32.09	22.70	1.91	1.79	40.49	98.9	$(Co_{1.72}Cu_{1.13}Fe_{0.11}Ni_{0.10})S_4$
2	32.75	21.95	1.47	1.94	41.07	99.1	$(Co_{2.19}Cu_{1.13}Fe_{0.10}Ni_{0.10})S_4$
3	32.46	22.03	1.57	1.89	40.71	98.6	$(Co_{1.88}Cu_{1.19}Fe_{0.10}Ni_{0.11})S_4$
		Average of 3 analyses: =					$(Co_{1.88}Cu_{1.19}Fe_{0.10}Ni_{0.11})S_4$

Native Bismuth

Two very small grains of native bismuth have been observed during recent ore microscopic examination of Creta ores. The native bismuth was recognized by its yellowish white color, very low hardness (softer than anilite), very high reflectance (about 60%), and weak anisotropism under oil immersion. One of the grains occurs as an irregularly shaped grain that is present as an inclusion in anilite (Fig. 10e). Figure 10f shows the results of a SEM-EDS analysis of that grain. The strong bismuth peaks and the lack of peaks for other elements confirms the optical determination of this mineral. The very low copper peak is the result of the small size of the bismuth grain and the presence of the enclosing chalcocite. The other occurs as a free bismuth grain in shale.

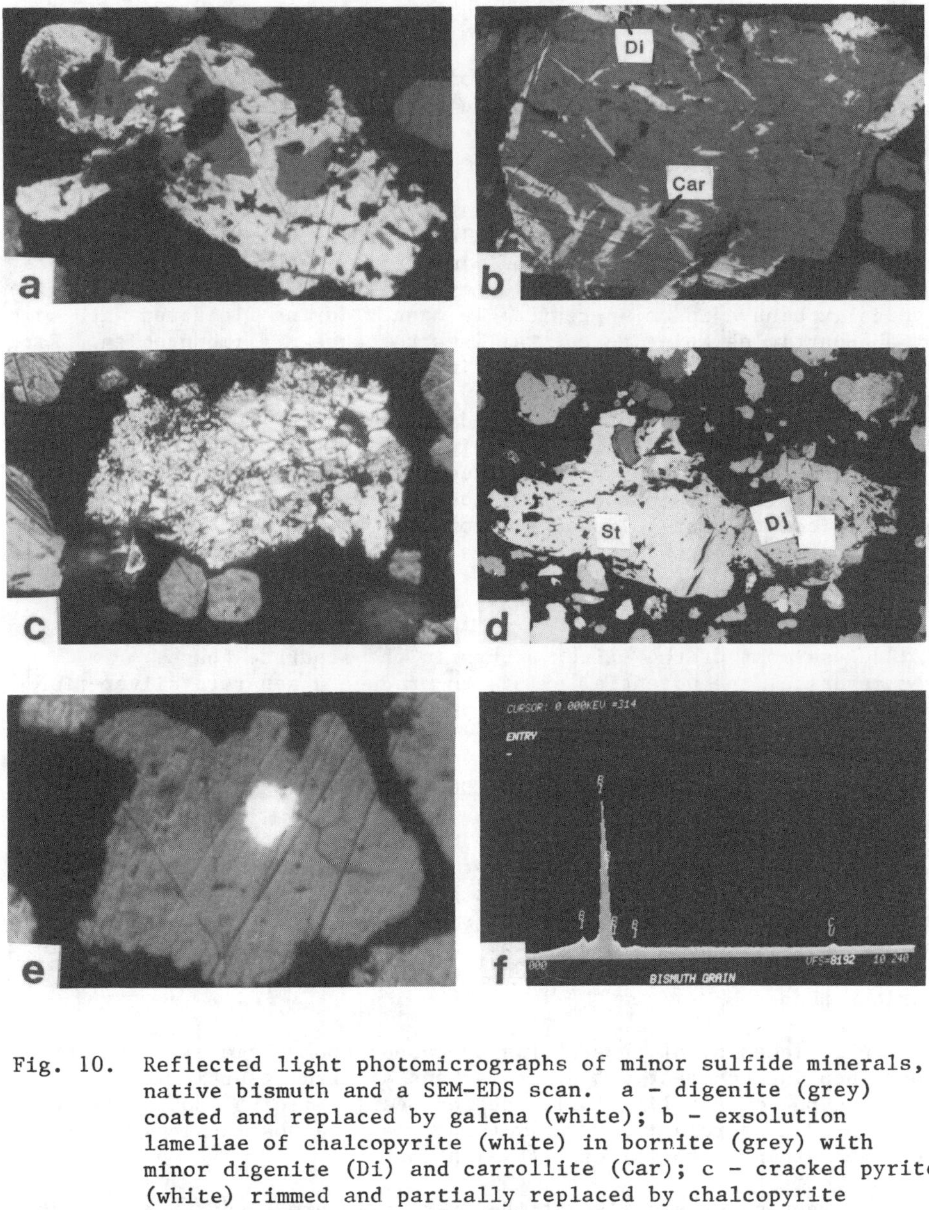

Fig. 10. Reflected light photomicrographs of minor sulfide minerals, native bismuth and a SEM-EDS scan. a - digenite (grey) coated and replaced by galena (white); b - exsolution lamellae of chalcopyrite (white) in bornite (grey) with minor digenite (Di) and carrollite (Car); c - cracked pyrite (white) rimmed and partially replaced by chalcopyrite (grey); d - stromeyerite (St) intergrown with argentiferous djurleite (Dj); e - native bismuth (white) included in copper sulfide; f - SEM-EDS scan for the bismuth grain. Grains a,b,c,d are about 100 μm long, the stromeyerite-bearing grain is 280 μm long, and the bismuth grain is 6 μm across.

APPLICATIONS OF ORE MICROSCOPY TO PROCESS MINERALOGY

The results of this ore microscopic and electron microprobe study have an important application to the beneficiation of this type of copper-silver ore as well as to a better understanding of their genesis. The treatment of the Creta ores by flotation resulted in a loss of about 25% of the copper in the heads to the tails. This ore microscopic examination of the Creta ores has shown three principal forms of copper occurrence that would be expected to cause copper losses during flotation. The most important of these is the presence of the platy crystals that are 10 µm and less. Particles of this size typically behave in an unpredictable manner during flotation with some of the particles going to the tailings to cause a slime problem. Part of the losses would be expected to involve binary locked copper sulfide-pyrite particles, especially for those ores that were derived from the upper portion of the orebody. Toward the lower portion of the orebody where the spores are only partially replaced, binary locked copper sulfide-spore particles might be expected to be a contributing factor in the copper losses. Ore microscopic study and particle counting of polished sections of the Creta tailings has shown that two-thirds of the copper losses are in the form of the 10 µm grains, and that one-third occurs as binary locked copper sulfide-pyrite and copper sulfide-spore particles (Hagni, 1978).

Because most of the silver occurs in the separate phase, stromeyerite, the potential exists to produce a separate silver-rich concentrate from the final Creta copper concentrate.

SUMMARY AND CONCLUSIONS

The copper sulfide grains in the copper ore at the Creta mine exhibit a wide variety of forms that have originated by epigenetic replacement processes. The largest grains were formed through the replacement of megaspores, and the smaller ones by the replacement of colloform pyrite bodies. Even the smallest grains were developed in large part through the replacement of tiny iron sulfide grains of various shapes.

The replacement of these forms by copper took place after the spore-bearing sediment was deposited and after diagenetic iron sulfides had crystallized. The copper-bearing fluids probably were introduced at a relatively early time, when the host shale was more permeable than it was after lithification.

The greenish-gray shales at the Creta mine are copper-bearing not because they contain syngenetic grains of copper sulfide but because they served as chemical traps, which caused copper to be deposited from very low temperature ore fluids, which permeated the favorable rocks. The most important chemical feature of the potential ore horizons was their high content of megaspores and their contained organic matter. Equally essential to the eventual deposition of copper was the diagenetic formation of iron sulfides. The abundant organic matter produced a reducing environment in the host shale during diagenesis and provided an energy source for sulfate-reducing bacteria, which in turn promoted the early formation of pyrite and pyrrhotite. Pyrite was most abundantly produced immediately beneath the gypsum cap, which prevented the penetration of oxygenated water from above. The presence of organic matter, which formed a reducing environment, and the availability of sulfur in a reduced state as pyrite and pyrrhotite formed the favorable conditions in which copper was trapped or precipitated from subsequently introduced fluids.

The ultimate origin of the introduced fluids is a subject for conjecture. The paleogeographic setting at Creta was one in which a sabhka environment may have existed, and it is possible that the copper-bearing fluids rose through the influence of evaporation in a Permian sabhka flat (Renfro, 1974; Smith, 1976).

ACKNOWLEDGMENTS

The cooperation of Eagle-Picher Industries in allowing for the sampling of the Creta mine is especially appreciated. The authors acknowledge SEM-EDS analyses by John Thompson and Auger microprobe analyses by Harlan Rice at the University of Missouri-Rolla, and electron microprobe analyses by Dan Kremser at Washington University, St. Louis, Missouri. The SEM and electron microprobe analyses were done as a part of a grant to one of us (RDH) from the United States Bureau of Mines.

REFERENCES

Arnold, R. G., 1969, "Pyrrhotite Phase Relations below 304 $\pm 6^{o}$C at <1 atm Total Pressure", Economic Geology, Vol. 64, pp. 405-419.

Berendsen, P., and Lambert, M. W., 1981, "Copper Sulfides in the Lower Permian Redbeds of South-Central Kansas", Part 1: Stratigraphy and Host Rock Lithology, Kansas Geological Survey Bulletin, Vol. 223, 19 pp.

Bernardini, G. P., Corsini, F., and Tanelli, G., 1973, "Djurleite from Calabona, Sardina," Mineralogical Magazine, Vol. 39, pp. 25-29.

Brockie, D. C., private communication, 1984.

Dingess, R. P., 1976, "Geology and Mining Operations at the Creta Copper Deposit of Eagle-Picher Industries, Inc.," Symposium on Stratiform Copper Deposits of the Midcontinent Region, K. S. Johnson and R. L. Croy, eds., Oklahoma Geological Survey Circular 77, pp. 15-24.

Dingess, R. P., 1976, private communication.

Goldhaber, M. B., and Kaplan, I. R., 1974, "The Sulfur Cycle," The Sea: Vol. 5, Marine Geochemistry, E. D. Goldberg, ed., Wiley-Interscience, pp. 589-655.

Hagni, R. D., "Ore Microscopy Applied to Beneficiation," Mining Engineering, Vol. 30, pp. 1437-1447.

Hagni, R. D., and Gann, D. E., 1976, "Microscopy of Copper Ore at the Creta Mine, Southwestern Oklahoma," Symposium on Stratiform Copper Deposits of the Midcontinent Region, K. S. Johnson and R. L. Croy, eds., Oklahoma Geological Survey Circular 77, pp. 40-50.

Johnson, K. S., 1976a, "Permian Copper Shales of Southwestern United States," Gisements Stratiformes et Provinces Cupriferes, P. Bartholome, ed., Soc. Geol. Belgique, pp. 383-393.

Johnson, K. S., 1976b, "Permian Copper Shales of Southwestern Oklahoma", Symposium on Stratiform Copper Deposits of the Midcontinent Region, K. S. Johnson and R. L. Croy, eds., Oklahoma Geological Survey Circular 77, pp. 3-14.

Johnson, K. S., and Brockie, B. C., 1973, "Eagle-Picher Industries, Inc's Copper Mine at Creta", Stop 10, Igneous Geology of the Wichita Mountains and Economic Geology of Permian Rocks in Southwest Oklahoma, Oklahoma Geological Survey, Guidebook for Field Trip No. 6, Geological Society of America Annual Meeting in Dallas, pp. 29-32.

Jones, W. R., "'Red Beds' Copper Deposits", 1965, in Mineral and Water Resources of New Mexico, New Mexico Bureau of Mines and Mineral Resources Bulletin 87, pp. 172-173.

Kidwell, A. L., and Bower, R. R., 1976, "Mineralogy and Microtextures of Sulfides in the Flowerpot Shale of Oklahoma and Texas," Symposium on Stratiform Copper Deposits of the Midcontinent Region, K. S. Kidwell and R. L. Croy, eds., Oklahoma Geological Survey Circular 77, pp. 51-60.

Lambert, Michael W., 1979, "Copper Sulfides in the Permian Redbeds of Kansas," Unpubl. M. A. thesis, Indiana University, 99 pp.

Lambert, M. W., Benendsen, P., and Ripley, E. M., 1981, Copper Sulfides in the Lower Permian Redbeds of South-Central Kansas, Part 2: Ore Mineralogy, Kansas Geological Survey Bulletin 223, 15 pp.

Lockwood, R. P., 1976, "Geochemistry and Petrology of Some Oklahoma Redbed Copper Occurrences," Symposium on Stratiform Copper Deposits of the Midcontinent Region, K. S. Kidwell and R. L. Croy, eds., Oklahoma Geological Survey Circular 77, pp. 61-68.

Morimoto, N., Koto, K., and Shimazaki, Y., 1969, "Anilite, Cu_7S_5, A New Mineral," American Mineralogist, Vol. 54, pp. 1256-1268.

Potter, R. W., II, 1977, "An Electrochemical Investigation of the System Copper-Sulfur", Economic Geology, Vol 72, pp. 1524-1542.

Potter, R. W., II, 1979, private communication.

Renfro, A. R., 1974, "Genesis of Evaporite-Associated Stratiform Metalliferous Deposits -- A Sabkha Process," Economic Geology, Vol. 69, pp. 33-45.

Ripley, E. M., 1984, private communication.

Ripley, E. M., Lambert, M. W., and Berendsen, P., 1980, Mineralogy and Paragenesis of Red-Bed Copper Mineralization in the Lower Permian of South Central Kansas", Economic Geology, Vol. 75, pp. 722-729.

Schouten, C., 1946, "Some Notes on Micropseudomorphism," Economic Geology, Vol. 41, pp. 348-382.

Skinner, B. J., "The System Cu-Ag-S," Economic Geology, Vol. 61, pp. 1-26.

Smith, E., 1976, "Sabkha and Tidal-Flat Facies Control of Stratiform Copper Deposits in North Texas," Handbook of Strata-Bound and Stratiform Ore Deposits, K. H. Wolfe, ed., Chapter 8, Vol. 6, pp. 407-446.

Williams, Neil, 1976, private communication.

Wilson, L. R., 1962, "Permian Plant Microfossils from the Flowerpot Formation, Greer County, Oklahoma," Oklahoma Geological Survey Circular 49, 50 pp.

Chapter 14

PRE-ORE POTASSIUM METASOMATISM, CREEDE MINING DISTRICT, COLORADO

by P. M. Bethke, R. O. Rye, and P. B. Barton, Jr.

U. S. Geological Survey, Reston, VA

ABSTRACT

Rhyolitic welded-tuff wallrocks of the epithermal base and precious metal veins of the Creede district were pervasively altered by the addition of more than two billion metric tons of potassium some 1.5-2 million years before mineralization. Sodium, calcium and magnesium were strongly depleted, yielding a nearly binary quartz + potassium feldspar assemblage containing as much as 13 weight percent K_2O. This large-scale metasomatism, originally noted by Steven and Rattle (1965), took place progressively by initial alteration of plagioclase phenocrysts to orthoclase or microcline followed by alteration of the groundmass feldspar to orthoclase and gradual change of the sanidine phenocrysts to more Or-rich compositions. Oxygen isotope and chemical studies show that the metasomatism resulted from the interaction of the tuffs with deeply circulating heated ground water and suggest that the potassium metasomatism of rhyolitic rocks is the facies equivalent of propylitization of volcanic rocks of more basic composition.

Chapter 15

STABILITY AND ASSOCIATIONS OF NATURAL TELLURIDES

by W. C. Kelly, E. J. Essene, and A. M. Affifi

University of Michigan
Ann Arbor, MI

ABSTRACT

Occurrences and associations of natural tellurides are constrained by the relative fugacities of Te_1, S_2, and O_2 in specific mineralizing environments. Some are rare (e.g., $FeTe_2$) and others absent (e.g. $MoTe_2$, ZnTe) in nature because high $f\text{-}Te_2/f\text{-}S_2$ ratios are rare or unattainable, whereas others (e.g., Ag_2Te, HgTe) require lower ratios and are common. Complex and seemingly chaotic sulfide-telluride associations are, in fact, highly systematic and depend both on temperature and fugacity ratios. Some older sulfides act as buffers via reactions such as $PbS + Te_2 = PbTe + S_2$, thereby affecting the deposition and paragenesis of younger tellurides and the combined or uncombined mode of occurrence of precious metals. Thermochemical calculations of telluride-sulfide-oxide equilibria can have pragmatic value. The thermochemical approach tells us little about the Te "source", but is beginning to unravel the systematics of telluride mineralogy and controls at the precipitational end of the line.

Chapter 16

CHARACTERIZATION OF TITANIUM DIBORIDE

by W. J. Croft and R. J. Arnott

Army Materials and Mechanics Research Center
Watertown, MA

ABSTRACT

A series of commercial titanium diboride powders have been characterized in terms of X-ray crystallography, composition, particle size distribution, and particle shape. These materials were used in sintering experiments from which the effect of impurities on the sintering temperature was observed. Transmission electron microscopic observations have been made on ion milled and chemically milled hot pressed material. Dislocations have been observed as well as grain boundary phases. Limited bonding between grains seems to account for poor mechanical properties. Observations have been made of preferred crystallographic orientation in hot pressed material.

Chapter 17

THE ZINC-IRON-MANGANESE ORE BODIES OF FRANKLIN AND STERLING HILL, SUSSEX COUNTY, NEW JERSEY

by R. W. Metsger

The New Jersey Zinc Company
Ogdensburg, NJ

ABSTRACT

The metamorphic zinc-iron-manganese ore in the Grenvillian rocks of the Reading Prong in Sussex County, N.J. is unique. The singularity of its mineral composition and associations has fueled controversy among economic geologists and mineralogists for more than a hundred years. The silicate-oxide ore, in two synformal concentrations has been mined since about 1838. The Franklin mine was depleted in 1954, while the one at Sterling Hill is still producing. Although the two mines are perhaps best known for their almost three hundred mineral species and have been a mecca for collectors, they -- as importantly -- have provided a window through which to observe more than a billion years of geologic history. When pieced together, studies reveal a sequence of probable sedimentary deposition, intense folding, pseudo-viscous flow with attendant negative diapirism, uplifting, erosion, karstification, hydrothermal alteration, burial, faulting, igneous intrusion, uplift, saprolitization, and glaciation.

MEMORIALS AND TESTIMONIALS

PAUL F. KERR MEMORIAL SYMPOSIUM DINNER

Andy H. Vassiliou - Introduction

Ladies and gentlemen, good evening and welcome to the Paul F. Kerr Memorial Symposium dinner. I believe we are all here to honor the memory of a great teacher and advisor, and a true friend -- Professor Kerr, or Dr. Kerr, or to most of us who were his students, "Pappy" Kerr.

Most of us who attended the Paul F. Kerr Memorial Symposium earlier today at the New York Hilton had the opportunity to read, or listen to, or discuss scientific papers presented in honor of Dr. Kerr. In addition, all of the speakers had the opportunity to reminisce about their association with Dr. Kerr, who was credited as having contributed substantially to their graduate education and the development of their careers.

Dr. Kerr's two daughters -- Ruth and Nancy -- who attended the Symposium at the Hilton are also here with us tonight. I think we should keep tonight's festivities on an informal basis since we are all old friends or former students of Dr. Kerr. Some of us may want to say a few words about our experiences at the time we were students at Columbia, and hopefully include some anecdotes relating to Professor Kerr. I would like to invite everyone to do so, but first we have arranged for two of us to be the lead speakers, whom I will introduce in a few moments.

We are of course at the Faculty House of Columbia, and I think we all know why we chose this place. It was Professor Kerr's favorite meeting place. I know I was here a few times -- Dr. Kerr brought us here for lunch or for some other festive occasion. So let's make it an evening of remininscing about the times when we worked with Dr. Kerr, and I would like to start by asking Nancy Kerr Del Grande to talk to us first.

Those of us who attended the Symposium earlier today met Nancy, a theoretical physicist, who gave one of the exciting papers at the Symposium relating to airborne temperature surveys and heat flow anomalies for exploration geology. Nancy Kerr.

Nancy Kerr Del Grande

Thank you. It's indeed a very great privilege and honor and pleasure being here at this dinner in honor of my dad. And now I speak for my sister and myself. This is a great privilege.

In the memorial dinner for our dad, Paul Kerr, I would like to share a few glimpses about Paul Kerr from the view of one of his daughters growing up as a little girl and coming here to the faculty club, maybe once every other week when my sister was studying medicine at P & S, and mother and I and dad would come here and we would meet all sorts of people, professors.

Dad had one son called Junior, who died at an early age. However, he had many adopted sons, and you are here tonight, many of you. And that's what makes it an extended family. This is dad's extended family and you are my brothers and sisters. So I have a number of brothers and sisters here, since many of you referred to him endearingly as "Pappy", and we called him dad.

At an early age I learned about sibling rivalry with my unknown brothers and sisters, for this was rivalry for my dad's time and attention. While he was out on the Colorado Plateau with his graduate students, my mother and my sister and I spent summers with my grandmother at Palo Alto, California. The names on my dad's itinerary included Grand Junction, Marysvale, Big Rock Candy Mountain. When I was eight years old I thought maybe dad would bring home some candy from Big Rock Candy Mountain. And then, of course, the names included Temple Mountain, Lisbon Valley and Grants, New Mexico.

Some thirty years later, I would study the use of remote sensing from a satellite to explore these areas for energy and mineral resources. Dad was a legend in the world of geology. My sister and I both went into different fields. She studied medicine and became the first woman neurosurgeon in the USA to be board certified in neurosurgery. I became the first woman physicist to design and conduct an underground experiment to test a nuclear weapon in Mercury, Nevada. Shortly before dad died, he said that the best investment he every made was in the education he gave his daughters. I know he also took great delight and pride in the successes of his former graduate students.

Schermerhorn Hall, Columbia University, was instrumental in my being a physicist. At thirteen, when I read Explaining the Atom by S. Hecht, I read about the first atomic pile at Columbia University. It was directly below my dad's desk at Schermerhorn. It's very fortunate that they moved it to Soldiers Field in Chicago before it went critical. I remember my dad mentioned that he had told the Fermi team he didn't think it would work. But I also remember as a little girl, there was a man who stopped me on the street, on 120th street, and said, "What are they doing over there at Schermerhorn?" And I don't know whether he was a spy from the enemy or whether he was part of the FBI. But I didn't know what was going on, other than they were modernizing; they were making it a nice new modern building and were also doing some interior decorating. And so, when I read three years later in Hecht's book what was really going on...

You know, he really was a very ideal security person, because his family didn't know.

I was a little bit annoyed that all this interesting atomic activity was going on at Columbia, and that interested me in going to Los Alamos and becoming a physicist, and really trying to explain the atom, and I am still trying to explain the atom. I hope maybe in the next year that I will be able to explain the atom in a new region, of energy region, where much of theoretical physics has made great improvements recently, in the last three or four years. And the experimentalists with synchrotrons are likely to be able to almost make enough improvements to keep the theoreticians honest. Anyway, I am interested in that area.

So Schermerhorn and the situation here influenced me to aspire to be a physicist and to work at Los Alamos. I ended up, after four years of graduate work at Stanford University, at the Lawrence Livermore Laboratory in 1959. Fifteen years later, with my dad's encouragement, I transferred to technology developed for nuclear testing...actively mapping the earth's temperature for geoexploration, after geothermal resources and a little bit of finding hot water to do geochemistry for mineral phase analysis. Recently, I formed a company, Geotemp Corporation, for this purpose. But I'll just be consulting, and my main full-time job is with Lawrence Livermore Laboratory. I've already put in twenty-five years, and I tend to continue another fifteen years.

Also during the 1970's my sister studied geology at Arizona State University and purchased four gold mines in Arizona. She and dad shared some good times in connection with this endeavor, for the few years before he died. So, he was patient. He had two daughters that went off in completely different fields, we thought. But they weren't so different: I was doing x-rays and diffraction and satellite geoexploration, and...he kept quiet because that wasn't too far (I found out today) from many things he was doing. And then she (my sister) was studying geology in the 70's. So he finally, being patient, saw his daughters sort of drift back to geology.

You will recognize my dad, Paul Kerr, as I read his two favorite poems. One is "If", by Kipling, the other is in a book by Robert Schuller.

If you can keep your head when all about you are losing
theirs and blaming it on you,
If you can trust yourself when all men doubt you, and make
allowance for their doubting too,
If you can wait and not be tired by waiting,
or being lied about don't deal in lies,
or being hated, don't give way to hating,
and yet don't look too good nor talk too wise,

If you can dream and not make dreams your master,
If you can think and not make thoughts your aim,
If you can meet with triumph and disaster and treat those
two imposters just the same,
If you can bear to hear the truth you've spoken, quested
by knaves to make a trap for fools,
or watch the things you gave your life to, broken,
and stooped and build them up with worn out tools
If you can make a heap of all your winnings and risk it
on one pitch of turn and toss,
and lose and start again a few beginnings and never
breath a word about your loss,
If you can force your heart and nerve and sinew to serve
your turn long after they are gone,
and to hold on when there is nothing holding to you
except the will which says to you hold on,
If you can talk into crowds and keep your virtue or walk
with kings nor lose the common touch,
If neither foes nor loving friends can hurt you,
If all men count with you, but none too much,
If you can build the unforgiving minute with sixty seconds
worth of distance run,
Then yours is the earth and everything that's in it,
And what is more, you'll be a man my son.

I am certain that Paul Kerr inspired many of his sons here in this room.

This final short poem so expresses Paul Kerr's character:

When faced with a mountain I will not quit,
I will keep on striving until I climb over,
Find a pass through, tunnel underneath,
Or simply stay and turn the mountain into
A gold mine with God's help.

Paul Kerr turned to God personally as his Lord and Savior, as well as reverently as the great mineralogist and creator.

Andy H. Vassiliou

I would now like to introduce our next scheduled speaker, former student and close associate of Professor Kerr, Dr. Arthur M. Langer. Art received his MA degree (in Petrology) from Columbia in 1960 and his Ph.D. (in Mineralogy) in 1965. He has been employed since then at the Mount Sinai Medical Center, where he is the Associate Director of the Environmental Science Laboratory, and he is also an Associate Professor at the Medical School of the Center. I believe Art will share some of his experiences when he was a student at Columbia working under the advisement of Dr. Kerr.

Art Langer

I would like to tell you something, first to Kerr's daughters. Pappy was special...he was a super person, he really was. He was kind, he was understanding, he was everything that a Columbia professor should be. He was good to us as graduate students, he was caring and he was everything that a good professor should be. I had a few words that I wanted to say about Pappy Kerr. I don't regard Professor Kerr as Paul Kerr; he is Pappy. I think all of us know him as Pappy Kerr.

I come from Pappy's Cenozoic era. The Cenozoic is the most recent, and I am the last of Pappy's students at Columbia before his retirement. I am the last in his tenure position, so to speak. And Art Rohl and I laughed about this. We said, "Well, Cenozoic." I don't feel Cenzoic. I am under a lot of pressure, I take a lot of heat, and I am physically and chemically, I guess, referring to archaean gneiss. We all go through those changes. But, of course, the memory of Papp Kerr remains with us, and when we talk about Pappy, we smile to ourselves.

He was really quite a guy, and I guess that I was the last in-resident Ph.D. My appointment was at the Mount Sinai Hospital and I was allowed to use the facilities at Mount Sinai where we studied silicosis, asbestosis, and so on. And Pappy would come into my office (my office was the mineral museum, 420 Schermerhorn, which is now part of the Department of Psychology or Art), and he would sit down and we would talk. I can't remember very many amusing times as a graduate student, but I can certainly remember those times when Pappy came into my office. We talked about a lot of things.

I remember very well the first time Pappy came into my office and we talked about one of my colleagues who was to receive a post with a prestigious geochemical group in the United States. And I said, "Professor Kerr, I don't understand this...this person doesn't have the personality to integrate with all those people. How could this person get this appointment?"

"Well, Art, I guess every organization needs someone like this." I said, "What's that?" He said, "It's a son of a bitch."

Now, I don't know whether you ever heard your father use profanities, but I can tell you that I have been a bad influence on many people, and I must have been a bad influence on Pappy Kerr. But he and I got along very well.

I heard stories about your father. You won't believe this, he was a nonconformist back in the old days. This is way before my time. This was told to me by Kurt Lowe who was in the Geology Department at Columbia at the time your father was an Assistant Professor, maybe an instructor, and Douglas Johnson was chairman of the Department. At that time the Journal Club looked after the cultural development of the graduate students, and Douglas Johnson used to read poetry to the graduate students and the faculty. And young Paul Kerr would sit in the back of the room, near a table on which there was a radio. While Johnson read poety, Paul Kerr would turn on the radio. Of course, the tubes of the radio had to be warmed up in those days. Pappy Kerry would turn on the radio and turn it on full. And then, just as it was about to turn on, he would turn it down. Kurt Loew says that everyone in the room used to focus on Pappy Kerr. So you see, he was a nonconformist. Your father was different. He broke the mold.

There are a few other stories, but I don't think I am gonna go through them all. I want you to know that your father's name also creeps up in the medical literature. Those of us doing work in the field find references of Burke and Kerr, who studied the mineral contents of lungs. And I always derived a great deal of pleasure seeing his name in the medical literature.

One day I went into Pappy Kerr's office and said: "Professor Kerr, what should I do? I've been offered this wonderful appointment down at the Mount Sinai School of Medicine and I have this appointment with the astrology branch of the U.S. Geological Survey. What do you think I should do?" He said, "Well, how much do they pay?" I said, "Well, I have to tell you this, the Mount Sinai School of Medicine certainly pays a lot more." He said, "I would then take it." With that good advice, I sent a letter down to Phil Bethke (of the U.S.G.S.) and I said I would take a one year leave of absence, and that was twenty years ago.

Well, what's there more to say. I liked Pappy, I really did. I liked him immensely. I think he was decent to all of us. When I speak for myself, I speak for my contemporaries. He was absolutely super. And I remember (and this absolutely stands out) how he used to say, "Look, you have to speak for yourself, you can't let your rights sit under a bushel basket. You got to get out there so that people may see you are what you say you are." I think that he taught us (my contemporaries) that if we are not for ourselves, who is for us? And if we are only for ourselves, then what are we to do? And that is stupendous!

Nancy Kerr

Dad wouldn't let me use a swear word, and I had to be dressed properly. And he seemed so perfect it drove me crazy sometimes. Once he drove me to Mt. Holyoke, and he had a bad cold, something like bronchitis. He was pretty ill and it was pouring outside and it was windy, and after we went three miles on the turnpike we had a flat tire. And he said, "Damn." And I said, "Oh, boy, finally human!"

Harry Dahl

Nancy, nobody is perfect. Your father was very close. I think I express the view of all of his students: absolutely fantastic! There was no other person that supported his graduate students the way your father did...an absolutely fantastic person...

Ronald J. Arnott

Actually, at one time when we were there he had fifteen graduate students. He supported all of them.

Otto C. Kopp

This will give you some idea of the way Pappy treated his graduate students, back in the good old days when we taught each other how to do things -- how to run the x-ray equipment and all the rest -- it was very typical.

One day the x-ray fluorescence unit stopped working. So Bill Bassett and I took it upon ourselves to fix this instrument. If you ever read the Phillips manual you would be impressed by the lack of detail. In fact, the manual said, "Loosen the knob," and in fact, there were two knobs. Unfortunately, Bill and I chose the wrong one. And as we loosened the knob, out came the insulating oil. Bill called Phillips and they told him that there was no way to repair that thing in the department; we had to send it back, and it would probably cost around ninety dollars. But that was back in the fifties...and that was like salary for a graduate student. And I could remember the two of us walking into Pappy's office and telling him what had happened. And to the best of my recollection, he said, "If you never do anything, you will never do anything wrong." And that was it.

When they sent it back to the factory, they told him he couldn't use it anymore. He had to buy a new tube. "I wonder what that costs!" I think that's just the way he was. He could have yelled at us or made us feel terrible, but that was not the way he acted. I guess he felt that that sort of thing was going to happen and we accepted that.

Bill Croft

In the years right after World War II, most of Professor Kerr's students were relatively older. They had done their undergraduate work before they went to war. Many of us had four or five years to make up, and we were very conscientious.

There is a kind of a story that is wonderful. I more or less had finished with my experimental work, and I think it was really time for me to get going, before Professor Kerr did what I thought was wonderful. He called me in one day and said, "It's about time you got your thesis done." What he actually said in effect was, "From now on, on pay day, I expect at least three or four pages per pay check." It was a wonderful way of getting me out of there. It was time to support my family.

Don Hausen

Dr. Kerr was always so conscientious involving students' time and money and capabilities. He was all the time looking after your personal interests. I will never forget when I first came to discuss with him my coming back to school to get my Ph.D. He looked at my credits, and all my credentials, my foreign language credits, and then he had me make a list of courses out of the catalog to satisfy the requirements for Ph.D. at Columbia. Just like a contract. I prepared it and sent it to him and he got it approved by the school. If I finished this course work, then I would be ready to start my thesis. That conscientious...he was that much concerned about his students.

After that, then things really started to work. "Have you got the first chapter finished? Where are you now?" He was very concerned about each individual student...giving them time limits...getting them through. As for the committees, the orals and the defense. He would go over the committee members that I would select..."That guy is pretty good...he is a primadonna, stay away from that guy." He knew his committee members, his people, very well. And he recognized that some of them would give his students a hard time. Very considerate. He had wisdom. He was a humanitarian. He understood that his students needed a special support system...they needed jobs. A very fine man. We certainly appreciated our mentor, "Pappy" Kerr.

Mrs. Don Hausen

I was not his student, I was his collaborator. I had to push Don. I have several memories of Dr. Kerr. One was when he stood up after Don had laid his whole future on the line on a paper that argued about the hydrothermal nature of uranium deposits at Temple Mountain. And this fine gentleman stood up and said, "It's one of the finest pieces of research I have ever seen," and I breathed. I hadn't been breathing for several weeks and it was getting hard. Then Don was to come back to Columbia and start working for his doctorate.

He would take us here to dinner. He would be very courteous because not many of his graduate students had ten children, and we did have ten children. It was a little handicap at times. The one thing I remember most of all is when he would go over Don's writing and send the material back. Don would bring it back and give it to me, I would retype it with the corrections, Don would look it over and correct it, Kerr would look it over and correct it and send it back, and I would type it again. On the seventh typing, I was typing what I had started the first time, and I said, "That's it!"

The day came when Don stood up to get his degree and ten little children and the wife stood beside him. And Dr. Kerr showed his colors that day. He asked us if we wouldn't mind posing for a photographer. Of course we would do anything for Dr. Kerr, he was such a nice gentleman. It was several days later our doorbell and telephone rang repeatedly in response to a front page picture of our family that had been sent via Associated Press throughout the country. At least they spelled our name right.

I also remember going to dinner with Dr. Kerr. He would take his wife by the arm; he was so courtly and gentlemenly to her. He never called her Helen in front of me (I didn't know her name until they told me today it was Helen). He always called her, always spoke to her as Mrs. Kerr. The courtliness of this gentlemen, the fine attitude he had towards all women, was one reason, especially to me, that he was the eminent Dr. Kerr.

Vivian Gornitz

I would like to say a few words. If Art Langer is in the Cenozoic, then I lay claim to the Holocene. I think I am the last of Kerr's students, and I'll say that my graduate years at Columbia are some of the happiest years in my life. And I felt really like I was being part of one big family, and I really remember that with great fondness.

I recall also my first trip out West, when I was heading for the Grand Canyon. This involves you too, Andy. I was coming out of Albuquerque, and we all met at Grants to visit some of the uranium mines. And I remember this adventure where I hardly got off the bus when Dr. Kerr had me and the others climbing up this mesa to chop down a tree. I think he was interested in age dating by counting the tree rings in the oldest tree that was imbedded in it. So we had to go out and pick that tree and chop it down and bring it back. And I still remember that.

Otto C. Kopp

One last comment. This concerns Peggy Hamilton who was Dr. Kerr's assistant. I shared an office with her for three years while I was there, and it was shortly after I left that I learned of her sudden and untimely death. I am sure that if she were alive today she would be one of the participants here.

Bill Bassett

I have a story to tell about Peggy Hamilton. I was not making good progress on my thesis and Dr. Kerr was expecting a report from me. I had to see him in his office. When Peggy saw me hesitating to go into his office she said, "It's a very gray, gloomy day. Why don't you wait until the sun is out and then go and see him." I took her advice, and when I went there on a sunny day, it wasn't bad at all."

Nancy Kerr

You certainly couldn't have had a nicer memorial. When we learned just about a month ago that this was going to happen, it really was an answer to a lot of things that we had wanted to start (in memory of our dad). We are certainly very happy to be here. I know he would be pleased.

Memorial of Paul Francis Kerr, January 12, 1897 - February 27, 1981

Otto C. Kopp
Department of Geological Sciences, University of
Tennessee, Knoxville, TN 37996-1410

Paul Francis Kerr passed away on February 27, 1981, after a brief illness. His death came just one day after his Keynote Address, "Reminiscences in Applied Mineralogy", had been presented at the AIME Symposium on "Process Mineralogy - Extractive Metallurgy, Mineral Exploration, Energy Resources", held in Chicago from February 22 to 26, 1981.

He was born on January 12, 1897, in Hemet, California, at the base of the San Jacinto Range along the San Jacinto fault of the San Andreas fault system. With savings from work in citrus orchards and bean fields, he entered Occidental College, receiving his BS in 1919. He received his Ph.D. from Stanford University in 1923. Occidental College conferred on him the Honorary Degree of Doctor of Science in 1960. He taught for a year at Stanford before assuming a teaching position at Columbia University in 1924. In that same year he married Helen Richardson Squire in the Serene Memorial Chapel at Stanford. They had three children, Paul Squire Kerr (who died as a young man), Ruth Elizabeth Kerr Jakoby, of Phoenix, Arizona, and Nancy Ann Kerr Del Grande, of San Leandro, California. His beloved wife, Helen, passed away on September 13, 1978.

The years at Columbia were busy and productive. It would be difficult to list all of his many accomplishments and honors during the period from 1924 to 1965; however, a partial list may serve to indicate the scope and breadth of his activities. During his tenure at Columbia, he sheparded 113 students to their graduate degrees and he produced 226 publications (including several books and monographs). He was Department Chairman at Columbia during the period 1944 to 1950 and the first Newberry Professor of Mineralogy from 1959 to 1965. After his retirement, he continued as Newberry Professor Emeritus, guiding his remaining graduate students to their degrees. He was instrumental in securing the property for, and naming, the Lamont Geological Observatory. Paul served as Secretary of the Mineralogical Society of America from 1934 to 1944, and was its President in 1946. He served as the Vice-President of the Geological Society of America in 1947. He received the third K. C. Li Medal in 1957 for his studies of tungsten minerals and the third Distinguished Member Award from the Clay Minerals Society, in 1970. Great Britain's Mineralogical Society made Paul on Honorary Member in 1972. He was a member of Phi Beta Kappa. He received many other awards and honors, and he served the geological community well throughout his life. An enlightening account of his activities was prepared by Ralph J. Holmes, Marjorie Hooker, Waldemar T. Schaller, Clarence S. Ross, and Donald M. Fraser at the time of his retirement in 1965. Many of his former students and friends contributed to that special Festschrift Volume of The American Mineralogist published in 1965 (Mineral Studies, Dedicated to Paul Francis Kerr, Volume 50, number 10). That volume contains a list of his publications through the time of his retirement in 1965. This Memorial completes Paul's list of publications by adding those published since his retirement.

He and Helen returned to California, where he continued to be active as Consulting Professor at Stanford University until 1977. During that period he gave seminars, assisted graduate students in mineralogical research, revised his well-known optical mineralogy text, and maintained an active research program on clay minerals and landslides.

We remember Paul as a pioneer in the application of mineralogical techniques. He is considered to be one of the fathers of applied mineralogy. His publications span a range of topics that would make most teachers and many researchers envious. They include studies on the X-ray diffraction process, clay minerals, uranium mineralization, quartz, tungsten minerals, differential thermal analysis, gems, alteration minerals, etc. Many members of the profession may have had their first contact with him through his text, Optical Mineralogy (1977), originally co-authored with A. F. Rogers, his mentor at Stanford, in 1933.

Those of us who were his students, his friends and colleagues will remember him as a patient teacher, a wise and helpful counselor, and a good friend. He expected much, but he gave much in return. Paul's genuine concern for generations of students earned him the affectionate nickname, "Pappy". We felt as close as his family, as if we were his adopted sons and daughters.

Some aspire to high places and personal honors, caring little about what happens to those they pass along the way. Paul Kerr was one of those special people who was more concerned about others than about his own esteem. In his quiet, unobtrusive way he was always there providing support and encouragement when it was most needed.

Paul Francis Kerr lived a long and productive life. Still, his death came as a shock. He is sadly missed by his family, his former students, and his many friends.

Selected Bibliography of Paul F. Kerr[1]

(with G. H. Megrue) Alteration of sandstone pipes, Laguna, New Mexico: Geol. Soc. Amer. Bull., v. 76, p. 1347-1360 (1965).

(with K. Omori) Infrared studies of saline sulfate minerals: Reply: Geol. Soc. Amer. Bull., v. 76, p. 283-295 (1965).

(with D. M. Davidson, Jr.) Uranium deposits at Kane Creek, Utah: Transactions of Soc. of Mining Engineers of AIME, v. 235, p. 127-132 (1966).

(with N. Guven) Selected Great Basin playa clays: Amer. Mineral., v. 51, p. 1056-1067 (1966).

(with N. Guven) Weathering effects on the structures of mica-type clay minerals: Amer. Mineral., v. 51, p. 858-874 (1966).

(with A. M. Langer) Mojave playa crusts: physical properties and mineral content: Jour. of Sed. Petrology, v. 36, p. 377-396 (1966).

(with J. T. Nash) Geological limitations on the age of uranium deposits in the Jackpile sandstone, New Mexico: Econ. Geol. v. 61, p. 1283-1287 (1966).

(with A. M. Langer) Evaluation of kaolinite and quartz differential thermal curves with a new high temperature cell: Amer. Mineral., v. 52, p. 509-523 (1967).

(with T. P. Rooney) Mineralogic nature and origin of phosphorite, Beaufort County, North Carolina: Geol. Soc. of Amer. Bull., v. 78, p. 731-748 (1967).

The Marysvale, Utah, uranium deposits: in J. D. Ridge, Ed., Ore Deposits of the United States, 1933-1963, Graton-Sales Volume, v. II, p. 1020-1042 (1968).

(with D. M. Davidson, Jr.) Uranium-bearing veins in plateau strata, Kane Creek, Utah: Geol. Soc. of Amer. Bull., v. 79, p. 1503-1526 (1968).

(and I. M. Drew) Recent quick-clay studies, 5. Quick-clay slides in the U.S.A.: Engineering Geol., v. 2, p. 215-238 (1968).

(with A. T. Gavasci) Uranium emplacement at Garnet Ridge, Arizona: Econ. Geol., v. 63, p. 859-876 (1968).

(with D. M. Hausen) Fine gold occurrence at Carlin, Nevada: in J. D. Ridge, Ed., Ore Deposits of the United States, 1933-1963, Graton-Sales Volume, v. I, p. 908-940 (1968).

(with G. H. Megrue) Alteration of sandstone pipes, Laguna, New Mexico: Reply: Geol. Soc. of Amer. Bull., v. 79, p. 791-794 (1968).

(with J. T. Neal and A. M. Langer) Giant desiccation polygons of Great Basin playas: Geol. Soc. Amer. Bull., v. 79, p. 69-90 (1968).

(and I. M. Drew) Clay mobility, Portuguese Bend, California: Calif. Div. Mines and Geol., Spec. Rept. No. 100, p. 3-16 (1969).

(and I. M. Drew and D. S. Richardson) Mud volcano clay, Trinidad, West Indies: Amer. Assoc. Pet. Geol. Bull., v. 54, p. 2101-2110 (1970).

(with V. Gornitz) Uranium mineralization and alteration, Orphan Mine, Grand Canyon, Arizona: Econ. Geol., v. 65, p. 751-768 (1970).

(with D. M. Hausen) X-ray diffraction methods of evaluating potassium silicate alteration in porphyry mineralization: in R. W. Boyle, Ed., Geochemical Exploration, Canadian Institute of Mining and Metallurgy Special Volume 11, p. 334-340 (1971).

(and R. A. Stroud and I. M. Drew) Clay mobility in landslides, Ventura, California: Amer. Assoc. Pet. Geol. Bull., v. 55, p. 267-291 (1971).

(and I. M. Drew) Clay mobility in Ridge Route landslides, Castaic, California: Am. Assoc. Pet. Geol. Bull., v. 56, p. 2168-2184 (1972).

(with A. Haji-Vassiliou) Uranium-organic matter association at La Bajada, New Mexico: Econ. Geol., v. 67, p. 41-54 (1972).

(with A. Haji-Vassiliou) Analytic data on nature of urano-organic deposits: Am. Assoc. Pet. Geol. Bull., v. 57, p. 1291-1296 (1973).

(and A. T. Gavesci and D. R. Bowes) Welded glass-breccias from Marysvale, Utah: Geol. Mag., v. 111, p. 15-22 (1974).

Quick clay: in F. Betz, Ed., Benchmark Papers in Geology - Environmental Geology, v. 25, p. 214-224 (1975).

Optical Mineralogy, 4th ed., 492 p., McGraw-Hill, Book Company, New york (1977).

Quick clays and other slide-forming clays: Engineering Geol., v. 14, p. 173-181 (1979).

Reminiscences in applied mineralogy: in D. M. Hausen, Ed., The Metallurgical Society Symposium Volume - Process Mineralogy in Extractive Metallurgy, Mineral Exploration, Energy Resources, p. 3-22 (1981).

[1]This bibliography includes all of Paul Kerr's publications since his retirement in 1965. A complete list of his earlier publications is included in Volume 50, number 10 of The American Mineralogist (1965), p. 1532-1545.

Acknowledgement. The author wishes to express his appreciation to the many persons who contributed to the preparation of this memorial, especially Paul's daughters, Ruth and Nancy, Phil Bethke, Don Hausen, Marty Molloy and Ben Page.

Excerpts from a Memorial to
Paul Francis Kerr, 1897 - 1981

Philip M. Bethke

U.S. Geological Survey
Reston, Virginia 22092

Paul F. Kerr - Age 38

Paul Francis Kerr, Newberry Professor of Mineralogy Emeritus at Columbia University and a former vice-president of the Geological Society of America, died on February 27, 1981, at Stanford University Hospital, Palo Alto, California, after a heart attack. With Paul F. Kerr's death, the profession lost one of its most successful educator-administrator-scholars, and his former students lost their most steadfast teacher-counselor-mentor-friend.

Paul Kerr entered Occidental College in the fall of 1915. He financed the later part of his college career by an assistantship in chemistry. Although his studies in chemistry and mathematics were interrupted by a brief period of military service, he graduated with his class in June 1919. The following fall he entered Stanford University with an assistantship to Professor Theodore Hoover, brother of the President. Although he began his studies in mining engineering, he soon switched to geology and came under the tutelage of Professor Austin Flint Rogers. Rogers introduced him to Professor D. L. Webster of the Physics Department, with whom he constructed a multiple X-ray diffraction unit, which he proceeded to use in work for his doctoral dissertation entitled "The Determination of Opaque Ore Minerals by X-ray Diffraction Patterns." This work, coming only a few years after the demonstrations by Debye and Scherrer in Germany and by Hull in America of the usefulness of the powder diffraction method, was one of the pioneering applications of the technique to problems of mineral identification.

Upon receiving his Ph.D. in 1923, Kerr replaced Rogers for the fall term at Stanford while Rogers was on leave. Kerr was offered a temporary position as lecturer in mineralogy at Columbia for the spring term and moved to New York to begin what was to be a 41-year association. Shortly after Kerr's arrival at Columbia, Professor Luquer, whom he was assisting, suffered a severe heart attack, and Kerr took over the responsibilities of the mineralogy program; shortly thereafter, he became a permanent member of the faculty, rising to the rank of full professor by 1940.

After his retirement from Columbia in 1965, the Kerrs returned to California where he was a consulting professor at Stanford until 1977. He remained active in research, advising graduate students and consulting until the time of his death. The final paper in his bibliography, "Reminiscences in Applied Mineralogy," was presented for him by D. M. Hausen as the keynote address to the AIME Symposium on "Process Mineralogy; Extractive Metallurgy, Mineral Exploration, Energy Resources" in Chicago just one day before his death. His final illness, coming just a few days before the symposium, prevented him from presenting the paper himself.

Paul Kerr's scholarly interests had as their underlying theme the application of mineralogy to problems of geology, engineering, and industrial processing. Indeed, his many contributions in that area led D. M. Hausen and W. C. Petruk to refer to him, in an editors' footnote to his final paper, as the "Father of Applied Mineralogy in this country." Although Paul Kerr's researches were as broad as mineralogy itself, three main areas particularly captured his attention early and held it for the duration of his career: clay mineralogy, the mineralogy of ore deposits, and the development of mineralogical techniques.

Kerr's fascination with clay minerals began early in his career when C. S. Ross sent him about a dozen samples on which to test the usefulness of X-ray diffraction techniques in clay mineralogy. Kerr's results confirmed Ross's conclusions based on optical studies and thus began a long and fruitful association between the two. Their collaborative studies provided the first firm bases for distinction between clay-mineral species and culminated in the first modern classification of clay minerals. It was natural that Paul Kerr should assume the leadership of the American Petroleum Institute's Project 49: Clay Minerals Standards program, which provided a set of clay-mineral standard samples from a large number of "type" localities. For each locality, a set of physical, chemical, optical, X-ray, and other data was collected by 23 specialists working in 10 different laboratories.

Paul Kerr's interest in clay minerals was not limited to systematic mineralogy, however, but extended to the application of clay mineralogy to the solution of a variety of geologic problems. He was especially concerned with (1) the "quick clays" and their role in landslides and slope stability, an interest that occupied him particularly during the last years of his career, and (2) the clay mineralogy of hydrothermal alteration haloes, an application to one of his other great loves, ore deposits. The classic alteration studies of the Santa Rita and Silver Bell porphyry copper deposits by Kerr and his students formed only a part of Kerr's contribution to the understanding of the genesis of ore deposits. His studies of tungsten mineralization in the western United States, which began in the mid-1930s with a comprehensive investigation at Mill City, Nevada, culminated in Memoir 15 of the Geological Society of America, Tungsten Mineralization in the United States, which was published in 1946. The study of the uranium mineralization at Marysvale, Utah, by Kerr and his students was published in 1957 as Special Paper 64 of the Society. Kerr's interest in uranium mineralization dated back to the early days of the Manhattan Project, when the interest in the availability of raw materials caused him to be sent to the Katanga district of what was then the Belgian Congo, where, upon his recommendation, and that of Philip Merritt, the famous Shinkolobwe Mine was reopened. Similar missions sent him to the Eldorado Mine at Great Bear Lake in the Northwest Territories, Canada, and to many localities in the United States. A long association with the U.S. Atomic Energy Commission followed, leading to a series of studies by Kerr and his students on various aspects of uranium mineralization in and around the Colorado Plateau. Kerr's involvement with uranium was only in part as a researcher. In 1945 he was selected by the Carnegie Endowment for International Peace to chair a commission charged with investigating problems associated with the international inspection of fissionable materials, and in 1955 he set up, on behalf of the United Nations, a program on raw materials for the First International Conference on the Peaceful Uses of Atomic Energy, held in Geneva, and edited the volume resulting from that program.

The techniques of mineral analysis were one of Paul Kerr's major interests, starting with his thesis project on the application of X-ray diffraction techniques to the identification of opaque minerals. His textbook, written with his mentor Rogers, was published in 1933 as Thin-Section Mineralogy; the revised edition was retitled Optical Mineralogy and was published in 1942. The third and fourth editions were extensively revised and published under Kerr's name alone in 1959 and 1977, respectively. Generations of students, many of them now retired, were introduced to the use of the petrographic microscope by this text, and it still continues to sell at the rate of well over a thousand copies per year! Although differential thermal analysis (DTA) had been applied to clay minerals as far back as 1913 and was widely utilized in the 1940s, the systematic studies of Kerr and his co-workers, along with those of Ralph Grim and of Toshio Sudo, gained acceptance of the method in clay-mineral studies. Kerr and his students made significant contributions to the development and improvement of DTA instrumentation. With J. Lawrence Kulp, he designed a multiple DTA unit in which a number of samples could be analyzed simultaneously under the same heating program, a technique that proved particularly useful in the study of clays. With Otto C. Kopp, he adapted the apparatus to provide for the analysis of materials, such as sulfides, which yielded corrosive decomposition products, an adaptation which he and J. A. Dunne later improved. Always on the lookout for new methods that could be applied to mineralogical problems, Kerr was among the first to implement such techniques as infrared and ultraviolet spectroscopy and X-ray fluorescence in his laboratory.

Kerr's contributions to systematic mineralogy included his efforts with C. S. Ross in straightening out the nomenclature of clay mineralogy, the first description (with others) of the new minerals alleghanyite, cattierite, dickite, hydrotungstite, sengierite, tungomelane, umohoite and vaesite, and the publication of a prodigious amount of definitive data on a wide variety of minerals.

It is simply not possible to document herein all the applications of mineralogy to which Paul Kerr contributed. He had a continuing interest in many other areas including gems and gemology, the so-called spark plug minerals, such as dumortierite, and the mineralogy of saline deposits. His bibliography boasts more than 250 entries, only the most important of which are listed in the following additional Bibliography. A complete bibliography through 1965, compiled by Marjorie Hooker, is published in the American Mineralogist, volume 50, no. 10, 1965, p. 1532-1545, and is updated in Otto Kopp's Memorial in this volume from American Mineralogist, volume 69. As impressive as Paul Kerr's contributions to science through teaching and research are, they tell only part of the story. His exceptional abilities as an administrator naturally brought to him many responsibilities, both at Columbia and within several scientific societies. Kerr took over as chairman of the Department of Geology upon the illness and retirement of Douglas Johnson in 1942; he retained the position until 1950. He was largely responsible for the acquisition by Columbia of Torrey Cliff, the estate of Thomas W. Lamont, on which site the Lamont-Doherty Geological Observatory now stands. His additional service to Columbia is documented in the minutes of the innumerable committees on which he served and which he often chaired, as vice-president of the American Association for the Advancement of Science; and as chairman of that society's Section of Geography and Geology. He also served as chairman of the Section of Geology of the New York Academy of Sciences and as president of the Kappa Chapter of the Society of the Sigma Xi, of the Faculty Club of Columbia University, and of the New York Mineralogical Club.

Paul Kerr received many honors and awards in recognition of his outstanding contributions. He was a member of Phi Beta Kappa and the Society of the Sigma Xi. In 1957 he was awarded the K. C. Li medal for his contributions to the geology and mineralogy of tungsten deposits, and he was awarded an honorary doctorate by his alma mater, Occidental College, in 1960. He was the third recipient of the Distinguished Member Award of the Clay Minerals Society in 1970, and in 1972 was elected an honorary member of the

Mineralogical Society of Great Britain. Perhaps the recognition that pleased him most was the assembly of more than 250 colleagues, former students, and friends at a reception and dinner in his honor on the occasion of his retirement from Columbia in 1965. At that time, he was presented a manuscript copy of the festschrift volume of the American Mineralogist (volume 50, no. 10, October 1965) dedicated to him by his friends and former students. In that volume, compiled and edited through the efforts of the late Ralph J. Holmes and William J. Croft, tributes by Holmes, Donald Fraser, Marjorie Hooker, Clarence S. Ross, and Waldemar Schaller gave special insights into Paul Kerr's life and career. They have been a fruitful source of information for this memorial.

Paul Kerr's contributions to our science have been enormous. All of us are in his debt, but none so much as his former students; most of us owe our careers to him, and all of us were the beneficiaries of his obvious and deep-felt concern for our well-being and our progress. This concern resulted in generations of students referring to him as "Pappy," a nickname that he knew had its roots in respect and affection, and one that bespoke most eloquently of the unique and special relationship that he established with his students.

Additional Bibliography of P. F. Kerr

The determination of opaque ore minerals by x-ray diffraction patterns: Economic Geology, v. 19, p. 1-34 (1924).

A simple rotation apparatus: American Mineralogist, v. 9, p. 169-171 (1924).

(and Schenck, H. G.) Active thrust faults in San Benito County, California: Geological Society of America Bulletin, v. 36, p. 465-494 (1925).

(and Cabeen, C. K.) Electrical conductivity of ore minerals: Economic Geology, v. 20, p. 729-737 (1925).

The significance of strain structure in quartz from Ducktown, Tennessee: American Mineralogist, v. 11, p. 206-209 (1926).

Significance of the Matilija overturn: Geological Society of America Bulletin, v. 39, p. 1087-1102 (1928).

An artificial gemstone isomorphous with spinel: American Mineralogist, v. 14, p. 259-264 (1929).

(with Ross, C. S.) Dickite, a kaolin mineral: American Mineralogist, v. 15, p. 34-39 (1930).

(with Ross, C. S.) The kaolin minerals: American Ceramic Society Journal, v. 13, p. 151-160 (1930).

(with Ross, C. S.) The clay minerals and their identity: Journal of Sedimentary Petrology, v. 1, p. 55-65 (1931).

Bentonite from Ventura, California: Economic Geology, v. 26, p. 153-168 (1931).

(with Ross, C. S.) The kaolin minerals: U.S. Geological Survey Professional Paper 165, p. 151-176 (1931).

(with Ross, C. S.) The manganese minerals from a vein near Bald Knob, North Carolina: American Mineralogist, v. 17, p. 1-18 (1932).

Montmorillonite or smectite as constituents of fuller's earth and bentonite: American Mineralogist, v. 17, p. 192-198 (1932).

The occurrence of andalusite and related minerals at White Mountain, California: Economic Geology, v. 27, p. 614-643 (1932).

(with Rogers, A. F.) Thin-section mineralogy: New York, McGraw-Hill Book Co., 311 p. (1933).

Geology of the tungsten deposits near Mill City, Nevada: University of Nevada Bulletin 28, no. 2, 46 p. (1934).

(with Ross, C. S.) Halloysite and allophane: U.S. Geological Survey Professional Paper 185, p. 135-148 (1934).

(and Jenney, P.) The dumortierite-andalusite mineralization at Oreana, Nevada: Economic Geology, v. 30, p. 287-300 (1935).

U-galena and uraninite in Bedford, New York cyrtolite: American Mineralogist, v. 20, p. 443-450 (1935).

(with Bray, R. H., and Grim, R. E.) Application of clay mineral technique to Illinois clay and shale: Geological Society of America Bulletin, v. 46, p. 1909-1926 (1935).

(and Cameron, E. N.) Fuller's earth of bentonitic origin from Tehachapi, Calif.: American Mineralogist, v. 21, p. 230-237 (1936).

The tungsten mineralization at Silver Dyke, Nevada: University of Nevada Bulletin 30, no. 5, 67 p. (1936).

Attapulgus clay: American Mineralogist, v. 22, p. 534-550 (1937).

(with Burke, H. E.) The nature of mineral particles in sputum and ash of the lungs of silicotics: Journal of Industrial Hygiene and Toxicology, v. 20, p. 535-555 (1938).

Tungsten mineralization at Oreana, Nevada: Economic Geology, v. 33, p. 390-427 (1938).

A decade of research on the nature of clay: American Ceramic Society Journal, v. 21, p. 267-286 (1938).

1940 A pinitized tuff of ceramic importance: American Ceramic Society Journal, v. 23, p. 65-71 (1940).

Tungsten-bearing manganese deposit at Golconda, Nevada: Geological Society of America Bulletin, v. 51, p. 1359-1389 (1940).

(with Rogers, A. F.) Optical mineralogy (2nd edition): New York, McGraw-Hill Book Co., 390 p. (1941).

(and Erichsen, A. I.) Origin of the quartz deposit at Fazenda Pacu, Brazil: American Mineralogist, v. 27, p. 487-499 (1942).

(and Young, F.) Hydrotungstite, a new mineral from Oruro, Bolivia: American Mineralogist, v. 29, p. 192-310 (1944).

(and Holmes, R. J.) X-ray study of the tantalum mineral simpsonite: Geological Society of America Bulletin, v. 56, p. 479-504 (1945).

Cattierite and vaesite; new Co-Ni minerals from the Belgian Congo: American Mineralogist, v. 30, p. 483-497 (1945).

(and Holmes, R. J., and Knox, M. S.) Lattice constants in the pyrite goup: American Mineralogist, v. 30, 498-504 (1945).

Tungsten mineralization in the United States: Geological Society of

America Memoir 15, 241 p. (1946).
Kaolinite after beryl from Alto do Giz, Brazil: American Mineralogist, v. 31, p. 435-442 (1946).
Alteration studies: American Mineralogist, v. 32, p. 158-162. Presidential Address, 27th Annual Meeting, Mineralogical Society of America, 1946 (1947).
(and Kulp, J. L.) Differential thermal analysis of siderite: American Mineralogist, v. 32, p. 678-680 (1947).
(and Kulp, J. L.) Multiple differential thermal analysis: American Mineralogist, v. 33, p. 387-419 (1948).
(with Vaes, J. F.) Sengierite; a preliminary description: American Mineralogist, v. 34, p. 109-120 (1949).
(with Kulp, J. L.) Improved differential thermal analysis apparatus: American Mineralogist, v. 34, p. 839-845 (1949).
(and Kulp, J. L., Patterson, C. M., and Wright, R.) Hydrothermal alteration at Santa Rita, New Mexico: Geological Society of America Bulletin, v. 61, p. 275-347 (1950).
(and Graf, D. L.) Trace element studies, Santa Rita, New Mexico: Geological Society of America Bulletin, v. 61, p. 1023-1052 (1950).
(and Holland, H. D.) Differential thermal analysis of davidite: American Mineralogist, v. 36, p. 563-572 (1951).
(with Kulp, J. L., and Kent, P.) Thermal study of the Ca-Mg-Fe carbonate minerals: American Mineralogist, v. 36, p. 643-670 (1951).
Alteration features at Silver Bell, Arizona: Geological Society of America Bulletin, v. 62, p. 451-480 (1951).
(and Kulp, J. L.) Pre-Cambrian uraninite, Sunshine mine, Idaho: Science, v. 115, no. 2978, p. 86-88 (1952).
Hydrothermal alteration and weathering, in Crust of the earth-a symposium: Geological Society of America Special Paper 62, p. 525-543 (1955).
The natural occurrence of uranium and thorium: International Conference on the Peaceful Uses of Atomic Energy, Geneva, 1955, Proceedings, Geology of uranium and thorium, v. 6, p. 5-59 (1956).
Rock alteration criteria in the search for uranium: International Conference on the Peaceful Uses of Atomic Energy, Geneva, 1955, Proceedings, Geology of uranium and thorium, v. 6, p. 679-684 (1956).
(and Kelley, D. R.) Urano-organic ores of the San Rafael Swell, Utah: Economic Geology, v. 51, p. 386-391 (1956).
(and Brophy, G. P., Dahl, H. M., Green, J., and Woolard, L. E.) Marysvale, Utah uranium area-geology, volcanic relations and hydrothermal alteration: Geological Society of America Special Paper 64, 212 p. (1957).
(and Bodine, M. W., Jr., Kelley, D. R., and Keys, W. S.) Collapse features, Temple Mountain uranium area, Utah: Geological Society of America Bulletin, v. 68, p. 933-981 (1957).
(with Kopp, O. C.) Differential thermal analysis of sulfides and arsenides: American Mineralogist, v. 42, p. 445-454 (1957).

(with Kelley, D. R.) Clay mineralogy and ore, Temple Mountain, Utah: Geological Society of America Bulletin, v. 68, p. 1101-1116 (1957).

Criteria of hydrothermal replacement in Plateau uranium strata: Second United Nations International Conference on the Peaceful Uses of Atomic Energy, Geneva, 1958, Proceedings, v. 2, Survey of raw materials resources, p. 330-334 (1958).

(and Hamilton, P. K.) Chrome mica-clay, Temple Mountain, Utah: American Mineralogist, v. 43, p. 34-47 (1958).

(with Kelley, D. R.) Urano-organic ores at Temple Mountain, Utah: Geological Society of America Bulletin, v. 69, p. 701-755 (1958).

Uranium emplacement in the Colorado Plateau: Geological Society of America Bulletin, v. 69, p. 1075-1111 (1958).

(with Kopp, O. C.) Differential thermal analysis of sphalerite: American Mineralogist, v. 43, p. 732-748 (1958).

(with Kopp, O. C.) Differential thermal analysis of pyrite and marcasite: American Mineralogist, v. 43, p. 1079-1097 (1958).

(with Kopp, O. C.) Differential thermal analysis of evaporites: American Mineralogist, v. 44, p. 674-679 (1959).

(with Hamilton, P. K.) Umohoite from Cameron, Arizona: American Mineralogist, v. 44, p. 1248-1260 (1959).

(with Huang, C. K.) Infrared study of the carbonate minerals: American Mineralogist, v. 45, p. 311-324 (1960).

(with Molloy, M. W.) X-ray spectrochemical analysis; an application to certain light elements in clay minerals and volcanic glass: American Mineralogist, v. 45, p. 911-936 (1960).

(with Dunne, J. A.) Differential thermal analysis of galena and clausthalite: American Mineralogist, v. 46, p. 1-11 (1961).

(with Barrington, J.) Uranium mineralization at the Midnite mine, Spokane, Washington: Economic Geology, v. 56, p. 241-258 (1961).

(with Abdel-Gawad, A. M.) Urano-organic mineral association: American Mineralogist, v. 46, p. 402-419 (1961).

(with Molloy, M. W.) Diffractometer patterns of A.P.I. reference clay minerals: American Mineralogist, v. 46, p. 583-605 (1961).

(with Bollin, E. M.) Differential thermal pyrosynthesis: American Mineralogist, v. 46, p. 823-858 (1961).

(and Barrington, J.) Clays of deep shale zone, Caillou Island, Louisiana: American Association of Petroleum Geologists Bulletin, v. 45, p. 1697-1712 (1961).

(with Barrington, J.) Breccia pipe near Cameron, Arizona: Geological Society of America Bulletin, v. 72, p. 1661-1674 (1961).

(with Barrington, J.) Alteration effects at Tuba Dike, Cameron, Arizona: Geological Society of America Bulletin, v. 73, p. 101-112 (1962).

(with Molloy, M. W.) Tushar uranium area, Marysvale, Utah: Geological Society of America Bulletin, v. 73, p. 211-235 (1962).

(with Adler, H. H.) Infrared study of aragonite and calcite: American Mineralogist, v. 47, p. 700-717 (1962).
(and Thomas, A. W., and Langer, A. M.) The nature and synthesis of ferrimolybdite: American Mineralogist, v. 48, p. 14-32 (1963).
(with Adler, H. H.) Infrared absorption frequency trends for anhydrous normal carbonates: American Mineralogist, v. 48, p. 124-137 (1963).
(with Abdel-Gawad, A. M.) Alteration of Chinle siltstone and uranium emplacement, Arizona and Utah: Geological Society of America Bulletin, v. 74, p. 23-46 (1963).
(with Bassett, W. A., Schaeffer, O. A., and Stoenner, R. W.) Potassium-argon dating of the late Tertiary volcanic rocks and mineralization of Marysvale, Utah: Geological Society of America Bulletin, v. 74, p. 213-220 (1963).
(with Omori, K.) Infrared studies of saline sulfate minerals: Geological Society of America Bulletin, v. 74, p. 709-734 (1963).
(with Adler, H. H.) Infrared spectra, symmetry and structure relations of some carbonate minerals: American Mineralogist, v. 48, p. 839-853 (1963).
(with Barrington, J.) Collapse features and silica plugs near Cameron, Arizona: Geological Society of America Bulletin, v. 74, p. 1237-1258 (1963).
(and Wilcox, J. T.) Structure and volcanism, Grants Ridge area: New Mexico Bureau of Mines and Mineral Resources Memoir 15, p. 205-213 (1963).
(and Jacobs, M. B.) Argillic alteration and uranium emplacement on the Colorado Plateau, in Clays and clay minerals: Twelfth National Conference on Clays and Clay Minerals, Proceedings, p. 111-128, London, Pergamon Press; New York, Macmillan Company (1964).
(with Adler, H. H.) Variations in infrared spectra, molecular symmetry and site symmetry of sulfate minerals: American Mineralogist, v. 50, p. 132-147 (1965).
(with Jacobs, M. B.) Hydrothermal alteration along the Lisbon Valley fault zone, San Juan County, Utah: Geological Society of America Bulletin, v. 76, p. 423-440 (1965).
(with Liebling, R. S.) Observations on quick clays: Geological Society of America Bulletin, v. 76, p. 853-878 (1965).
(with Geise, R. F.) The crystal structures of ordered and disordered cobaltite: American Mineralogist, v. 50, p. 1002-1014 (1965).

Memorial originally published by The Geological Society of America, 3300 Penrose Place, Boulder, Colorado. Reprinted with permission.

AUTHOR INDEX

SUBJECT INDEX

D

E

F

G

H

I

K

L

M

N

O

P

Q

R

S

T

U

V

X

Z